W9-CRX-684

PLEASE STAMP DATE DUE, BOTH BELOW AND ON CARD

WITHDRAWN
CALTECH LIBRARY SERVICES

WITHDRAWN

Physics and Chemistry in Space   Vol. 15

Edited by L. J. Lanzerotti, Murray Hill and
D. Stöffler, Münster

QB
603
A85
A87
1986

Sushil K. Atreya
III

Atmospheres and Ionospheres of the Outer Planets and Their Satellites

ASTR

With 90 Figures

Springer-Verlag
Berlin Heidelberg New York
London Paris Tokyo

Professor Sushil K. Atreya
Department of Atmospheric
and Oceanic Science
The University of Michigan
Space Research Building
Ann Arbor, MI 48109, USA

ISBN 3-540-16832-X Springer-Verlag Berlin Heidelberg New York
ISBN 0-387-16832-X Springer-Verlag New York Berlin Heidelberg

Library of Congress Cataloging-in-Publication Data. Atreya, S. K. Atmospheres and ionospheres of the outer planets and their satellites. (Physics and chemistry in space ; vol. 15) Bibliography: p. Includes index. 1. Outer planets – Atmospheres. 2. Outer planets – Ionospheres. I. Title. II. Series: Physics and chemistry in space ; v. 15. QC801.P46 vol. 15 530'.0919 s 86-26014 [QB603.A85] [551.5'0999'2]

This work is subject to copyright. All rights are reserved, whether the whole or part of the material is concerned, specifically those of translation, reprinting, re-use of illustrations, broadcasting, reproduction by photocopying machine or similar means, and storage in data banks. Under § 54 of the German Copyright Law, where copies are made for other than private use, a fee is payable to "Verwertungsgesellschaft Wort", Munich.

© Springer-Verlag Berlin Heidelberg 1986
Printed in Germany

The use of registered names, trademarks, etc. in this publication does not imply, even in the absence of a specific statement, that such names are exempt from the relevant protective laws and regulations and therefore free for general use.

Typesetting: Fotosatz GmbH, Beerfelden.
Offsetprinting and Bookbinding: Konrad Triltsch, Graphischer Betrieb, Würzburg.
2132/3130-543210

To My Parents,
Evelyn, and Chloé

Preface

One of the most fundamental discoveries of the solar system was the detection of four moons in orbit around Jupiter by Galileo Galilei in 1610. The discovery was significant not only in the context of Jupiter; it gave credence to and was instrumental in firmly establishing the heliocentric system of Nicolaus Copernicus. Almost four centuries after Galileo's discovery, extensive observations by the two Voyager spacecrafts have once again revolutionized our thinking about the major planets, their composition, structure, origin, and evolution. This book is an attempt at summarizing our present understanding of the atmospheres and ionospheres in the outer solar system, with particular emphasis on the relevant physics and chemistry. I was motivated to prepare this manuscript for the following reasons. First, after undergoing rapid expansion in the recent past, the subject has finally attained sufficient maturity to warrant a monograph of its own. Second, I have felt that as a result of recent observations, new and challenging problems have arisen whose resolution requires unconventional analysis and theoretical interpretation of existing data, as well as the collection of new kinds of data. I believe the time is ripe to put these issues in the appropriate scientific perspective, with the hope of stimulating novel theoretical, observational, and laboratory studies. I have highlighted the significant scientific problems throughout the book, especially at the end of each chapter.

The topics in this book were selected – after much discussion with colleagues and students – to cover essential aspects of the physical and chemical processes in the atmospheres and ionospheres of the outer planets and their satellites. Both measurements and theoretical concepts and formulations have been included. Because of space restrictions, I have found it necessary to limit the discussion of meteorology, magnetospheres, interior, and evolution to only those aspects that are clearly essential to the understanding of atmospheric and ionospheric phenomena. A background in physics, chemistry, and mathematics at the undergraduate level with some basic knowledge of meteorology, is adequate preparation for understanding the material in the book and for getting the most out of it.

The book is intended for researchers as well as graduate students in planetary science. Recognizing the obvious dichotomy between these two groups, I have attempted to maintain some balance in discussing the various topics. Graduate students and other new entrants to the field of planetary atmospheres should find the development of conceptual material and the formulations in the book most beneficial. On the other hand, seasoned researchers looking for new ideas and insights will be more interested in those parts of the book that deal with the critical evaluation of latest available information along with the extensive bibliography for tracking down the source. Because of the highly interdisciplinary nature of the field of planetary atmospheres, I have sometimes found it appropriate and convenient to develop the formulation of certain topics where they are discussed, rather than separately, e.g., all at the beginning of the book. The reader, however, should find the Subject Index most helpful for locating the desired formulation or topic, and for cross-referencing. Although I have been careful to present a balanced view of the subject, at times I have found it appropriate to advocate a certain viewpoint, or include a provocative suggestion. In all such instances of controversial ideas, caveats are discussed. It is my sincere hope that in its own way this book will help to inspire many people to pursue studies in planetary science and stimulate new research ideas with the ultimate goal of understanding the origin and evolution of planetary atmospheres.

Ann Arbor, August 1986   S. K. ATREYA

Acknowledgments

I wish to express my sincere appreciation to many colleagues: those who critically reviewed one or several chapters of the manuscript, those who supplied preprints of their work, and those with whom I have had stimulating discussions during the course of preparing the manuscript. The list of these individuals is long, the following, however, deserve special thanks: Jacques E. Blamont, Barney J. Conrath, Imke de Pater, Thomas M. Donahue, Von R. Eshleman, Daniel Gautier, Donald M. Hunten, Louis J. Lanzerotti, Glenn S. Orton, Tobias C. Owen, Ronald G. Prinn, Bill R. Sandel, Donald E. Shemansky, Edward C. Stone, Darrell F. Strobel, and G. Len Tyler. Parts of the manuscript were written in 1984 – 85 while I was on a sabbatical leave at Université de Paris (Pierre et Marie Curie). I am thankful to Claudie Blamont, whose tireless efforts helped to make my stay in Paris both productive and pleasant. Provocative inquiries from many of my former and present graduate students at Michigan and Paris – in particular, J. Hunter Waite, Paul N. Romani, Robert B. Kerr, and Jean Jacques Ponthieu – were helpful in clarifying some of the concepts discussed in the manuscript. Very special thanks are due to Deborah A. Swartz for her efficient and expert help in preparing several drafts of the manuscript. C. A. Rohn helped with proofreading and the bibliography. Much of my own research, which is reflected throughout the book, has been generously supported by the Planetary Atmospheres Program of NASA's Solar System Exploration Division.

S. K. ATREYA

Contents

1 Composition

It is generally believed that the planets in our solar system formed following the cooling and condensation of the primordial solar nebula, a large, dense cloud of dust and gas surrounding the juvenile Sun. The Sun and its planetary system represent perhaps only one of the thousands of fragments which may have resulted following the collapse of a large interstellar cloud. According to the above hypothesis of solar system formation, the constituents of planetary atmospheres would be in solar proportions provided that condensation, selective escape, fractionation, or evolution has not occurred. The massive major planets of our solar system — Jupiter, Saturn, Uranus, and Neptune — have large escape velocities which would prevent even the lightest of gases, hydrogen, from escaping. For example, the equatorial gravitational escape velocity (Appendix A1.1) at Jupiter is $57 \, km \, s^{-1}$, whereas the mean thermal velocity (Appendix A1.1) of hydrogen atoms at Jupiter's exospheric temperature (~ 1000 K) is only $4 \, km \, s^{-1}$. The time constant (also known as the e-folding time) for hydrogen atoms, i.e., the time required for escape of $\sim 70\%$ of the total atmospheric hydrogen, turns out to be 10^{300} years! Nonthermal escape is likely, but is not very significant at Jupiter and Saturn. On Uranus, nearly half of the hot hydrogen atoms produced on dissociation of molecular hydrogen by soft electrons are capable of overcoming the gravitational energy barrier (additional discussion of this phenomena on Uranus can be found in Chapter 2 on Thermal Structure and Chapter 4 on Vertical Mixing). For all practical purposes, however, the major planets are expected to preserve their original composition, unlike the terrestrial planets, whose atmospheres have undergone drastic evolution in the past 4.5 billion years. The bulk or gross composition of the major planets is expected to reflect the solar ratio of the elements. The best available information on the solar ratios of several elements, along with a list of constituents in which these elements are mainly incorporated on the major planets, is presented in Table 1.1.

Another model for the origin of the solar system, known as the nucleation model, is quite different from the one just described. It assumes that the cores of the giant planets formed first (Mizuno 1980; Mizuno et al. 1978). The scenario is as follows (Gautier and Owen 1985). A core consisting of refractory material, ices of water, methane, ammonia, clathrate-hydrates of CH_4, N_2, CO, etc. forms first. Once the core grows to a critical mass, it gravitationally traps the most volatile of gases (H_2, He, noble gases) which surround the core. The accretional heating of the core could also result in the vaporization of the

Table 1.1. Elemental ratios in a planetary atmosphere reflecting solar composition

Element	Constituent in the Jovian atmosphere[a]	Elemental ratios relative to H_2	
		(After Cameron 1982)	(After Cameron 1973)
H	H_2		
He	He	1.35×10^{-1}	1.39×10^{-1}
O	H_2O	1.38×10^{-3}	1.35×10^{-3}
C	CH_4	8.35×10^{-4}	7.42×10^{-4}
Ne	Ne (?)	1.95×10^{-4}	2.16×10^{-4}
N	NH_3	1.74×10^{-4}	2.35×10^{-4}
Mg		7.97×10^{-5}	6.67×10^{-5}
Si	SiH_4	7.52×10^{-5}	6.29×10^{-5}
Fe		6.77×10^{-5}	5.22×10^{-5}
S	H_2S (?)	3.76×10^{-5}	3.14×10^{-5}
Ar	Ar (?)	7.97×10^{-6}	7.37×10^{-6}
Al		6.39×10^{-6}	5.35×10^{-6}
Ca		4.70×10^{-6}	4.54×10^{-6}
Na		4.51×10^{-6}	3.77×10^{-6}
Ni		3.59×10^{-6}	3.02×10^{-6}
Cr		9.55×10^{-7}	7.99×10^{-7}
Mn		6.99×10^{-7}	5.85×10^{-7}
P	PH_3	4.89×10^{-7}	6.04×10^{-7}
Cl	HCl (?)	3.56×10^{-7}	3.58×10^{-7}
K		2.63×10^{-7}	2.64×10^{-7}
Ti		1.80×10^{-7}	1.75×10^{-7}
Co		1.65×10^{-7}	1.39×10^{-7}
Zn		9.47×10^{-8}	7.82×10^{-8}
F	HF (?)	5.86×10^{-8}	1.54×10^{-7}
Cu		4.06×10^{-8}	3.40×10^{-8}
V		1.91×10^{-8}	1.65×10^{-8}
Ge	GeH_4	8.80×10^{-9}	7.23×10^{-9}
Se	SeH_2	5.04×10^{-9}	4.23×10^{-9}
Li		4.51×10^{-9}	3.11×10^{-9}
Kr	Kr (?)	3.11×10^{-9}	2.94×10^{-9}
Ga		2.86×10^{-9}	3.02×10^{-9}
Sc		2.33×10^{-9}	2.20×10^{-9}
Sr		1.72×10^{-9}	1.69×10^{-9}
Zr		9.02×10^{-10}	1.76×10^{-9}
Br	HBr (?)	6.92×10^{-10}	8.49×10^{-10}
B	B_2H_6(?)	6.77×10^{-10}	2.20×10^{-8}
As	AsH_3 (?)	4.66×10^{-10}	4.15×10^{-10}
Rb		4.59×10^{-10}	3.70×10^{-10}
Xe	Xe (?)	4.39×10^{-10}	3.38×10^{-10}
Y		3.61×10^{-10}	3.02×10^{-10}
Mo		3.01×10^{-10}	2.52×10^{-10}

[a] Constituents marked "?" are expected to be present in the atmosphere; the compounds in the core are not listed.

various ices in the core, subsequently enhancing the atmospheric abundances of such volatiles as CH_4, NH_3, H_2O, etc. The nucleation model appears promising, especially for Uranus and Neptune, at whose heliocentric distances temperatures are cold enough to facilitate condensation of the various ices needed for the formation of the cores. Since not enough information is available concerning the thermochemical conditions and the accretional history of the core, one cannot confidently evaluate the abundances of volatiles in the atmospheres of the major planets based on the nucleation model. They could, however, differ substantially from solar proportions.

For the formation of Uranus and Neptune, yet another tantalizing scenario exists − that these planets are simply a conglomeration of millions of comets which were presumably orbiting the Sun at vast heliocentric distances (where Uranus and Neptune are now). If indeed the core of these planets does incorporate a substantial quantity of melted-down cometary material, the planetary density would be greater than that of a hydrogen-rich (Saturn-like) planet. This is because the comets are primarily composed of O, C, and N constituents. The densities of Uranus and Neptune are found to be much greater than that of Saturn. The comet-conglomeration hypothesis also implies that a substantial ocean of water should surround the rocky core, as water is the main ingredient of comets. According to the comet hypothesis, presence of at least some CO and N_2 in the atmosphere of Uranus and on some of its larger moons is likely. These constituents have so far not been detected in the Uranian system. Precise measurements of the atmospheric composition of the outer planets hold promise for discriminating between the various hypotheses for the formation of the giant planets, if not the solar system.

1.1 Bulk Atmospheric Composition

Prior to the Voyager observations, composition of the major planets was derived mainly by using infrared spectroscopic techniques on ground-based telescopes. The earth-based observations are capable of higher spectral resolution and routine temporal coverage. On the other hand, they suffer from poor spatial resolution and coverage, and contamination from the infrared spectrum of the Earth's atmosphere itself which is heavily sprinkled with H_2O and CO_2 absorptions. Nonetheless, terrestrial observations − especially those using high resolution Fourier Transform Spectroscopy on ground-based and airborne (e.g., Kuiper) observatories have provided a fairly reliable picture of the bulk composition of the atmospheres of major planets. These observations take advantage of the many "windows" in the Earth's infrared spectrum. They are located between 1 and 3 μm, around 5 μm, around 10 μm, and between 20 and 40 μm, as seen in Fig. 1.1 (Ridgway et al. 1976). Also shown in this figure is the infrared spectrum of Jupiter in the 1 to 40 μm range. The region below approximately 4 μm represents sunlight reflected from above the NH_3 clouds which are located in the 0.5 to 1 bar region at Jupiter, (additional

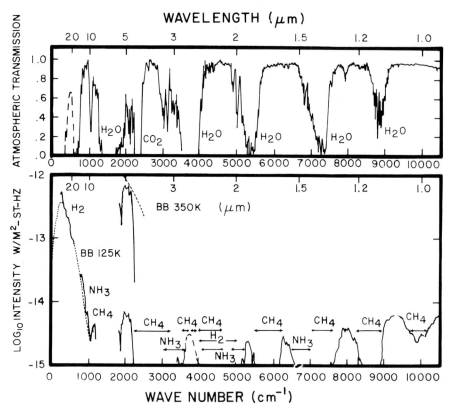

Fig. 1.1. *Top panel* Typical transmission of the Earth's atmosphere under good conditions, measured from Kitt Peak. Several "windows" in this transmission spectra are separated by strong telluric absorption bands at 1.14, 1.4, 1.9, 2.7 μm (all due to H_2O), 4.3 μm (CO_2) and 6.3 μm (H_2O). *Bottom panel* A composite of several ground-based infrared spectrum of Jupiter. For the 5 μm region, a "hot spot" spectrum (*upper curve* marked *Black Body* BB 350 K) is shown in addition to the one for the disk average brightness temperature (*lower curve* with BB 125 K). (After Ridgway et al. 1976)

details in Chap. 3 on Cloud Structure). Strong absorptions in CH_4 (3 v_3 band at 1.108 μm, etc.) and NH_3 are responsible for the shape of Jovian spectrum in this region. In addition, pressure-induced vibration-rotation spectra of H_2 (1 – 2.5 μm) and the H_2-quadrupole lines (0.6 – 1.1 μm) are also prominent in the region above 3000 cm^{-1} (H_2, being a homonuclear molecule, does not possess a permanent dipole moment; therefore its infrared spectroscopic signature at Jupiter, Saturn, etc. is the result of pressure-induced and quadrupole transitions).

Beyond approximately 4 μm, the spectra result mostly from thermal emissions. The pressure-induced rotational and translational spectrum of H_2 is responsible for the large infrared opacity in the 10 – 40 μm region; it also limits "visibility" down to the upper troposphere. Although the spectrum in this

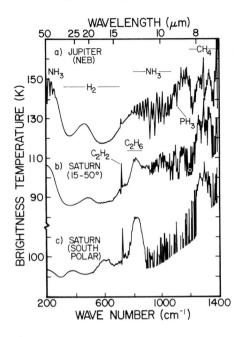

Fig. 1.2. Comparison of the Voyager infrared spectra of Jupiter's North Equatorial Belt and Saturn's low – mid latitude and the south polar regions. (After Hanel et al. 1981a)

region resembles Planck's radiation curve (for a black body radiation temperature of 125 K at Jupiter, Fig. 1.1), prominent spectral features of CH_4, NH_3, C_2H_2, C_2H_6, PH_3, etc. are also present. The majority of spectral features in this region are the result of stratospheric emissions, rather than absorptions, because of the existence of a strong thermal inversion at around the 100 – 200 mb pressure level.

A composite of typical spectra for Jupiter and Saturn obtained with Voyager Infrared Interferometric Spectrometer (IRIS) is shown in Fig. 1.2. For Jupiter, the spectra clearly show the high J rotational lines of NH_3 around 200 cm^{-1}. Their virtual absence at Saturn is indicative of the low saturation vapor pressure of NH_3 above the ammonia clouds. Other prominent features in these spectra are: broad pressure-induced translation-rotation transitions of H_2 [$S(0)$ and $S(1)$, between 300 and 700 cm^{-1}]; v_2-NH_3 rotation-vibration band (800 – 1200 cm^{-1}); v_4-CH_4 rotation-vibration band (1200 – 1400 cm^{-1}); C_2H_2 and C_2H_6 bands; and Q(1122 cm^{-1}) and R(1184 cm^{-1}) branches of PH_3 not masked by NH_3. Furthermore, at Saturn (and Titan), detection of C_3H_4 and C_3H_8 was made at respectively 633 cm^{-1} and 748 cm^{-1} (Fig. 1.3). Finally, Voyager 2 IRIS measurements (not shown) indicate a factor of 3 greater C_2H_6/C_2H_2 abundance ratio in the polar region than at lower latitudes of Jupiter (Hanel et al. 1979a, b). Such variation could be related to one of many factors: charged particle dissociation of CH_4 or other hydrocarbons in such a way as to give larger C_2H_6 than C_2H_2 production, equator to pole decrease in C_2H_2 (or increase in C_2H_6) abundance due either to transport or photochemical effects.

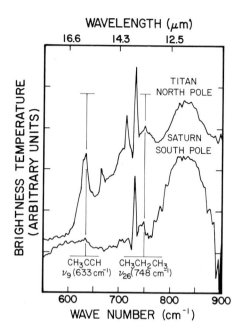

WAVELENGTH (μm)

Fig. 1.3. Tentative detection of methylacetylene (C_3H_4 or CH_3CCH) and propane (C_3H_8 or $CH_3CH_2CH_3$) from the Voyager IRIS spectra of Saturn and Titan. (After Hanel et al. 1981 a)

The 5 μm region is particularly suited for seeing deep into the troposphere, since both the terrestrial and the Jovian windows are nearly coincident in this frequency region. The terrestrial window extends from 4.3 μm (CO_2) to 6.3 μm (H_2O), whereas the Jovian window is from 4.5 μm (CH_4) to 5.2 μm (NH_3). It should therefore be possible to detect and measure the 5 μm flux originating deep in the troposphere, provided that breaks exist in the intervening cloud layers to permit its transmission. Observations for Jupiter indicate that the intensity at 5 μm is large only at a few "hot spots". One such observation of Jupiter at 5 μm indicates that the spectral formation occurred at a temperature of 300 K or greater; the corresponding pressure would be ≥ 5 bar. This same observation yielded an O/H ratio at Jupiter a factor of 1000 smaller than the solar O/H ratio (Larson et al. 1975). This single data point cannot be taken to represent globally averaged conditions at Jupiter, however. On the other hand, high spatial resolution observations done at 2.7 μm and 5 μm from Kuiper Airborne Observatory yield only a factor of 50 depletion in the Jovian O/H ratio at the 6-bar ($T = 288$ K) level (Bjorakar 1985). In-situ measurements of water vapor and clouds in the Jovian troposphere planned from the Galileo/Jupiter Probe will be highly instructive in resolving the important question of possible volatile depletions at Jupiter. At Saturn, the mean level of the origin of 5 μm flux is around 175 K (~ 2 bar pressure). Voyager infrared observations, however, yield data on Saturn hot spots with temperatures as high as ~ 210 K. Numerous minor constituents (H_2O, CH_3D, CO, GeH_4, HCN, PH_3) have been detected in the 5-μm region at Jupiter. Only CH_3D and PH_3 from this list have been definitely identified in the IR spectrum of Saturn.

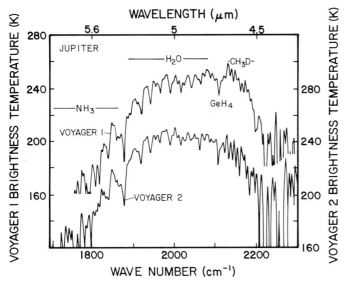

Fig. 1.4. The 5 μm region of Jupiter, as recorded by Voyager IRIS (Hanel et al. 1979b; copyright 1979 by the AAAS)

The detection and abundance of CO at Saturn (Noll 1985) should be regarded as tentative, as the spectroscopic information of the PH_3 lines − which are strongly mixed in with the CO lines − has not yet been included. The Voyager IRIS observations in the planetary 5-μm window at Jupiter are shown in Fig. 1.4. The NH_3 spectral features are prominent beyond the edge (5.2 μm) of this window. There is virtually no difference between the brightness or rotational temperatures of the Voyager 1 and 2 observations. The mole fractions of the trace constituents have profound implications for the convective and thermo-chemical processes in the deep atmosphere of Jupiter and Saturn, as discussed in Chapter 5 on Photochemistry and Chapter 3 on Cloud Structure.

The detection of helium had to await observations from flyby spacecraft, even though it is expected to be the second most abundant constituent of the major planets. The resonance emission of helium at 584 Å is relatively feeble. The earth-orbiting astrophysical observatories [such as Copernicus, International Ultraviolet Explorer (IUE), Hubble Space Telescope (HST), etc.] have been generally restricted to wavelengths above the Lyman-cutoff limit (i.e., $\lambda \geq 911$ Å), so that not even an attempt to detect the He 584 Å emission could be made from Earth orbit. Furthermore, it should be noted that the He 584 Å emission arises from the upper atmosphere, therefore it is incapable of pro-viding directly the helium abundance in the mixed atmosphere. The detection of helium in absorption has also not been possible, as He does not have a de-tectable absorption signature above ~ 500 Å. Its detection and abundance measurement are, therefore, based on the effect of $He - H_2$ collisions on the above-mentioned H_2-infrared absorptions, and on the atmospheric mean molecular weight. The former is studied using the spacecraft infrared spectrometers, while the latter is derived from the spacecraft radio occultation data.

As will be discussed in Chapter 2, the radio refractivity data yield information on the ratio of atmospheric temperature to the mean molecular weight (i.e., $T/\bar{m}$). With different assumed values of $\bar{m}$, different temperature profiles would be obtained. Based on these temperatures then, synthetic infrared spectra can be calculated for the region of H_2-absorption and compared with observations (the relevant spectral region is $200-600$ cm^{-1} at Saturn, and $280-600$ cm^{-1} at Jupiter due to the NH_3 absorption at Jupiter below 280 cm^{-1}). The best fit to the data fixes the value of the mean molecular weight. Since $\bar{m}$ is almost entirely composed of H_2 and He, its measurement, in effect, provides a determination of the helium mole fraction (Gautier et al. 1981). If abundances of constituents other than H_2 and He are substantial, their effect on $\bar{m}$ can be taken into account provided that their abundances are known. At Saturn, for example, CH_4/H_2 turns out to be $5 \times$ solar, thus contributing appreciably to the value of $\bar{m}$. The value of $\bar{m}$ is calculated using the following relation.

$$\bar{m} = 2\,q_{H_2} + 4\,q_{He} + 16\,q_{CH_4}\,,$$

where q's are the mole fractions. Knowing the CH_4 mixing ratio, along with the fact that

$$\sum_i q_i = 1, \quad \text{where} \quad i = H_2, He, CH_4\,,$$

one can still exploit the above-mentioned technique of determining the He/H_2 mixing ratio on the major planets. This technique, however, gives the value of the helium mole fraction only at the radio occultation points (generally 2, entry and exit).

The other method of determining the He/H_2 ratio takes advantage of the differential variation with frequency of the coefficients of H_2-H_2 and H_2-He collision-induced absorption (Gautier and Grossman 1972). This method uses only the infrared data. Such spectral dependence, along with the fact that the observed thermal emission spectrum is sensitive to both the He/H_2 ratio and the vertical temperature profile, permits a unique determination of q_{H_2} and the thermal profile (Conrath et al. 1984). Helium abundance obtained using this technique could potentially be erroneous because of lack of inclusion of infrared opacity due to any aerosols, and disequilibrium between ortho- and para-hydrogen states. Fortunately, both these effects are found to be of little consequence for the determination of He/H_2 ratio, at least at Jupiter and Saturn (Conrath et al. 1984).

The He/H_2 mixing ratios obtained in the above manner for Jupiter and Saturn are found to be quite different from one another. At Jupiter, the value is close to solar, within the range of statistical uncertainties. In Saturn's atmosphere, however, helium appears to be substantially depleted compared to its solar abundance. This finding may not necessarily imply differences in the bulk compositions of the two planets. In fact, it most likely means that in the interior of Saturn, helium is differentiated from molecular hydrogen and helium rain drops subsequently fall toward the core (Smoluchowski 1967). This process would not only deplete helium above the level of differentiation, it would also release gravitational energy. The latter is essential for explaining the strength of Saturn's internal heat source (further discussion will be presented in Chapter 2 on Thermal Structure).

The presently known composition of the atmospheres of Jupiter and Saturn is summarized in Tables 1.2 – 1.5. Some of the constituents listed in these tables (such as H, upper atmospheric hydrocarbons, etc.) are measured using

Table 1.2. Measured composition of Jupiter's atmosphere

Constituent	Spectral region [μm]	Mixing ratio[1] (relative to H_2)
H_2 (mol. hydrogen)	0.8, 0.64[2,3]	
	16.7 to 35.7	0.88 ± 0.036[4,5], relative to atmosphere
He (helium)	0.0584[6]	
	17, 28	$0.136^{+0.049}_{-0.045}$[4,5]
CH_4 (methane)	0.7, 1.1[7,8,9]	
	<7.41, >8.06	$(1.95 \pm 0.22) \times 10^{-3}$[10]
	0.125 to 0.16	$2.5^{+3}_{-2} \times 10^{-5}$ (1 μb region)[11]
$^{13}CH_4$	1.1	$^{12}C/^{13}C = 110 \pm 35$[12]
	7.4 to 8	$^{12}C/^{13}C = 160^{+40}_{-55}$[13]
C_2H_6 (ethane)	10[14]	
	12	5×10^{-6}[15]
	12	1.2×10^{-6} (above 100 mb level)[16]
	0.125 to 0.16	$2.5^{+2.0}_{-1.5} \times 10^{-6}$[11]
C_2H_2 (acetylene)	10[14]	
	13	$(2-8) \times 10^{-8}$ in the stratosphere[15]
	0.125 to 0.16	$<2.5 \times 10^{-6}$[11]
C_2H_4 (ethylene)	v_7 fundamental at 10.5	$7 \pm 3 \times 10^{-9}$[17] in north-polar region
C_3H_8 (propane)	13.4	$<6 \times 10^{-7}$[17] in north polar region
C_3H_4 (methylacetylene)	v_9 band at 15.8	$2.5^{+2}_{-1} \times 10^{-9}$[17] in north polar region; $<7 \times 10^{-10}$ at midlatitudes
C_4H_2 (diacetylene)	v_8 at 15.9	$<3 \times 10^{-10}$[17,18]
C_6H_6 (benzene)	v_{11} fundamental at 14.8	$2^{+2}_{-1} \times 10^{-9}$[17] in north polar region; $<2 \times 10^{-10}$ at midlatitudes[17]
CH_3D (deuterated methane)	5.0[19,20]	
	8.2 to 8.93; 4.55 to 4.76	$3.5^{+1.0}_{-1.3} \times 10^{-7}$[21] in North Equat. Belt (NEB)
HD (deuterated hydrogen)	0.7	$(1.02 \pm 0.14) \times 10^{-4}$[22], could be smaller

Table 1.2 (continued)

Constituent	Spectral region [μm]	Mixing ratio[1] (relative to H_2)
NH_3 (ammonia)	0.7, 0.645[7, 8] 8.3 to 9.1 9.1 to 11.8; 33.3 to 55.6	$(3.0 \pm 1.5) \times 10^{-4}$[23] in well mixed region 2×10^{-7} (at 0.2 bar level)[21] 1.5×10^{-5} (at 0.5 bar level)[21] 1.78×10^{-4} (at 1 bar level, assumed solar)[21]
PH_3 (phosphine)	10[24], 5[25, 26] 8.3 to 9.1	$(1-2) \times 10^{-7}$ in 0.2–0.6 bar region, and 6×10^{-7} in $p > 1$ bar[21, 27]
CO (carbon monoxide)	5.0	2×10^{-9}[28, 29, 30] in the stratosphere
GeH_4 (germane)	5.0[31] v_3 (4.74)	$(7 \pm 2) \times 10^{-10}$[21], NEB
H_2O (water)	5.0 4.76 to 5.26 2.7 and 5	1×10^{-6} ($T \geq 300$ K)[32] 1×10^{-6} at 2.5 bar, and 3×10^{-5} at 4 bar[21] Saturated for $P < 2$ bar, 4×10^{-6} in 2 to 4 bar region, 3×10^{-5} at 6 bar ($T = 288$ K)[33]
HCN (hydrogen cyanide)	5.0 13.5	1×10^{-7}[31] 2×10^{-9} in the stratosphere[34]
H (atomic hydrogen)	0.1216[6, 35]	Variable
H_2S (hydrogen sulfide)	2.7 0.16	$< 3.3 \times 10^{-8}$ in upper troposphere[36] $< 5.0 \times 10^{-9}$ in stratosphere[37]
$(H_2)_2$ (hydrogen dimer)	28.2[38, 39]	

[1] Mixing ratio (also called volume mixing ratio) refers to the ratio of constituent number density (or partial pressure) to the H_2-number density (or H_2-pressure). Only best current values given; [2] Kiess et al. (1960); [3] Spinrad and Trafton (1963); [4] Hanel et al. (1979a and b); [5] Gautier et al. (1981); [6] Judge and Carlson (1974); [7] Wildt (1932); [8] Kuiper (1952); [9] Owen (1965); [10] Gautier et al. (1982); [11] Festou et al. (1981); [12] Fox et al. (1972); [13] Courtin et al. (1983); [14] Ridgway et al. (1974); [15] Maguire W., personal communication (1982); [16] Kostiuk et al. (1983); [17] Kim et al. (1985); [18] Gladstone and Yung (1983); [19] Beer and Taylor (1973a); [20] Beer and Taylor (1973b); [21] Kunde et al. (1982); [22] Trauger et al. (1977); [23] Knacke et al. (1982); [24] Ridgway et al. (1976); [25] Larson et al. (1977); [26] Beer and Taylor (1979); [27] Encrenaz et al. (1978); [28] Beer (1975); [29] Beer and Taylor (1978); [30] Larson et al. (1978); [31] Larson et al. (1976); [32] Larson et al. (1975); [33] Bjoraker (1985); [34] Tokunaga et al. (1981); [35] Broadfoot et al. (1979); Clarke et al. (1980a); [36] Larson et al. (1984); [37] Owen et al. (1980); [38] Gautier et al. (1983); [39] Frommhold and Birnbaum (1984).

the stellar occultation or other techniques which are discussed separately in Chapter 2. Tables 1.3 and 1.5 list constituents whose abundances are not yet measured, but their existence suspected from theoretical or observational considerations. The bulk atmospheric composition of these planets is illustrated in cartoons shown in Fig. 1.5 and 1.6. These figures show roughly the levels

Table 1.3. Species expected, and their mixing ratios in a solar composition Jovian atmosphere

Constituent[a]	Mixing ratio (relative to H_2)	Principal isotopic ratios[b]
Ne (neon)	2.0×10^{-4}	$^{20}Ne/^{22}Ne = 8.2$ $^{20}Ne/^{21}Ne = 330$
H_2S (hydrogen sulfide)	3.8×10^{-5}	$^{32}S/^{34}S = 22.5$ $^{32}S/^{33}S = 125$
Ar (argon)	8.0×10^{-6}	$^{36}Ar/^{38}Ar = 5.3$ $^{36}Ar/^{40}Ar = 4.5 \times 10^{-6}$
HCl (hydrogen chloride)	3.6×10^{-7}	$^{35}Cl/^{37}Cl = 3.1$
N_2 (nitrogen)	$3 \times 10^{-6} - 10^{-9}$	$^{14}N/^{15}N = 272$
HF (hydrogen fluoride)	5.9×10^{-8}	
SeH_2 (selenium hydride)	5.0×10^{-9}	
Kr (krypton)	3.1×10^{-9}	$^{84}Kr/^{86}Kr = 3.3$ $^{84}Kr/^{83}Kr = 4.9$ $^{84}Kr/^{82}Kr = 4.9$ $^{84}Kr/^{80}Kr = 28$ $^{84}Kr/^{78}Kr = 161$
SiH_4 (silane)	2.0×10^{-9}, stratosphere 8×10^{-5}, $T > 1200$ K region	
GeS (germanium sulfide)	$\leq 10^{-9}$	
HBr (hydrogen bromide)	7×10^{-10}	$^{79}Br/^{81}Br = 1.02$
AsH_3 (arsane)	4.7×10^{-10}	
Xe (xenon)	4.4×10^{-10}	$^{132}Xe/^{131}Xe = 1.2$ $^{132}Xe/^{130}Xe = 6$ $^{132}Xe/^{129}Xe \simeq 1$ $^{132}Xe/^{128}Xe = 11.9$ $^{132}Xe/^{126}Xe = 225$ $^{132}Xe/^{134}Xe = 2.6$ $^{132}Xe/^{136}Xe = 3.1$

[a] In addition, mixing ratios of 10^{-9} to 10^{-12} are expected for the following species: CH_3CN, HI, SbH_3, CH_3NH_2, N_2H_4, COS, CH_3S, $(C_2H_5S)_2$, C_2H_4, C_3H_4, C_3H_8, C_4H_{10}, C_5H_{12}, C_6H_6, P_2, P_4, and B_2H_6. Some of these constituents have been detected at the polar latitudes.
[b] Also expected $^{16}O/^{18}O = 490$, $^4He/^3He = 5.6 \times 10^3$ (?).

of the cloud layers, and mixing ratios of volatiles through the lower stratosphere. The observed decrease with height in the mixing ratios of certain constituents, such as NH_3, H_2S, H_2O, etc. is due to photolysis or condensation. Whatever little is known about the compositions of Uranus and Neptune atmospheres is presented in Table 1.6. Note in this table the depletion of NH_3 and the enhancement of CH_4, but virtual absence of the heavier hydrocarbons. These issues will be discussed in Chapter 3 on Cloud Structure and Chapter 5 on Photochemistry. Even less is known about the atmosphere of

Table 1.4. Composition of the Saturn atmosphere

Constituent	Spectral region [μm]	Mole fraction (relative to H_2)
H_2	$S(0)$ and $S(1)$ quadrupole lines of the (4,0) and (3,0) rotational-vibrational system[1,2,3,4]	1.0 (but relative to atmosphere: 0.963 ± 0.024)[5]
He	17, 28	0.034 ± 0.028[5]
CH_4	3 ν_3 (1.1)[6,7,8,9]	
	ν_4 (7.7)	$4.5^{+2.4}_{-1.9} \times 10^{-3}$[10]
	0.125 – 0.16	1.5×10^{-4} (at 10^{-8} bar)[11]
$^{13}CH_4$	1.1	$^{12}C/^{13}C = 89^{+25}_{-18}$[9]
C_2H_6	ν_9 (12.2)[10,12,13]	$3.0 \pm 1.1 \times 10^{-6}$ N-hemisphere[10] $(3.1 \pm 0.7) \times 10^{-6}$ S-hemisphere[10]
C_2H_2	13.7	$2.1 \pm 1.4 \times 10^{-7}$ N-hemisphere $(P < 20$ to 50 mb)[10] $(0.5 \pm 0.1) \times 10^{-7}$ S-hemisphere[10]
C_3H_4	ν_9 (15.8)[14]	
C_3H_8	ν_{26} (13.4)[14]	
PH_3	$10^{(14,10)}$	$1.4 \pm 0.8 \times 10^{-6}$ $P > 3$ to 6 mb
	$10 – 11$[15] 5[6]	
CH_3D	5[10,16]	$3.9 \pm 2.5 \times 10^{-7}$[10]
CO	5	$(3 \pm 2) \times 10^{-9}$ tentative[17]
HD	P_4 (1) at 0.7467; and R_5 (0)[18,19]	$1.1 \pm 0.6 \times 10^{-4}$[19]
NH_3	$50, 5$[14,10]	$(0.5 – 2.0) \times 10^{-4}$ at 2.5 bar saturation at $P < 1.3$ bar
	0.6450[20] 1.56[21] Radio[22]	
H_2S	1.59[21] $0.21 – 0.25$[23]	Nondetection upper limit only
H	0.1216[24,25,26,27]	Variable
$(H_2)_2$	28.2[28,29]	

[1] Münch and Spinrad (1963); [2] Giver and Spinrad (1966); [3] Owen (1969); [4] Encrenaz and Owen (1973); [5] Conrath et al. (1984); [6] Trafton (1973); [7] Trafton and Macy (1975); [8] Lecacheux et al. (1976); [9] Combes et al. (1977); [10] Courtin et al. (1984); [11] Festou and Atreya (1982); [12] Gillett and Forrest (1974); [13] Tokunaga et al. (1975); [14] Hanel et al. (1981a); [15] Encrenaz et al. (1975); [16] Fink and Larson (1977); [17] Noll (1985); [18] Trauger et al. (1977); [19] Macy and Smith (1978); [20] Woodman et al. (1977); [21] Owen et al. (1977); [22] Gulkis et al. (1969); Gulkis and Poynter (1972); [23] Caldwell (1977); [24] Weiser et al. (1977); [25] Barker et al. (1980); [26] Clarke et al. (1981a); [27] Broadfoot et al. (1981a); [28] Gautier et al. (1982); [29] Frommhold and Birnbaum (1984).

Table 1.5. Upper limits of trace constituents on Saturn from IR spectra[a]

Molecule	Wavelength region [μm]	Upper limit [cm-amagat][b]	Reference
H_2O	5	15 (prec. μm H_2O)	1
HCN	3	0.1	1
	13.5	2.5×10^{-2}	2
H_2S	1.6	15	3
	1.6	10	4
	1.6	1	5
COS (carbonyl sulfide)	1.6	10	4
SiH_4	5	2.5×10^{-2}	1
GeH_4	5	2×10^{-3}	1
CH_3OH (methyl hydroxide)	1.6	10	4
CH_3NH_2 (methylamine)	1.6	2	4

[a] After Prinn et al. (1984).
[b] 1 cm-amagat = 2.687×10^{19} molecules cm^{-2}, also referred to as cm-am and cm-atm.
[1] Larson et al. (1980); [2] Tokunaga et al. (1981); [3] Martin (1975); [4] Fink and Larson (1979); [5] Owen et al. (1977).

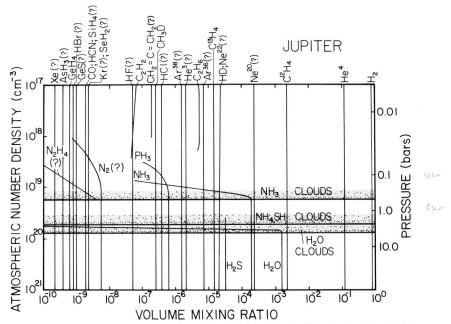

Fig. 1.5. A cartoon showing the locations of equilibrium clouds of H_2O, NH_4SH and NH_3, and the mixing ratios of constituents detected or expected (marked ?) in the lower atmosphere of Jupiter

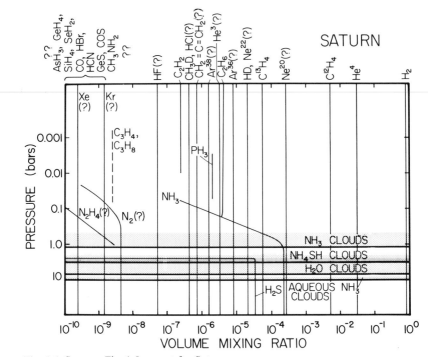

Fig. 1.6. Same as Fig. 1.5, except for Saturn

Pluto because of its vast heliocentric distance and the small size. At the subsolar temperature of Pluto, atmospheric methane would condense. From the
observations in the $1.5-2.5$ μm range it is concluded that the surface of Pluto
is covered with nearly pure methane ice (Soifer et al. 1980). Observations of
the 1.7 μm band also indicate absorptions characteristic of methane frost, and
perhaps CH_4 vapor (in equilibrium with the frost), (Cruikshank and Silvaggio
1978, 1980). Observations done between 6200 Å and 9000 Å place the CH_4 gas
abundance between 1 and 6.3 m-am (Martin 1975; Benner et al. 1978; Barker
et al. 1980). Since the escape velocity at Pluto is only $1-2$ km s^{-1}, most of the
gases would simply not be stable against blow-off. Only small quantities of
heavy volatiles (~ 0.3 cm-am CH_4; 155 m-am N_2, etc.; Trafton 1981; Hunten
and Watson 1982) can be retained in the atmosphere.

From Tables $1.2-1.5$ it can be concluded that within the range of measurement uncertainties, gases such as PH_3 and NH_3 are in solar proportions in
the mixed atmospheres of Jupiter and Saturn. The apparent depletion of NH_3
on Uranus and Neptune (Table 1.6) is most likely caused by the loss of this
constituent in an aqueous-NH_3 solution cloud which forms in the deeper
troposphere (see Chap. 3). As discussed earlier, the observed depletion of
helium on Saturn, and H_2O on Jupiter may not be representative of their bulk
compositions. Gases such as C_2H_2, C_2H_4, C_2H_6, GeH_4, CO, HCN are dis-

Table 1.6. Composition of Uranus and Neptune atmospheres

Constituent	Uranus	Neptune	Reference
H_2 (0.5 – 1.1 μm)	500 – 700 km-am	300 – 600 km-am	1
He	$He/H_2 = 0.18 \pm 0.07$		2, 3
HD [$R_5(0)$. $R_5(1)$, $P_4(1)$]	$1.8 \times 10^{-5} <$ $D/H < 4 \times 10^{-4}$	?	4, 5, 6, 7
NH_3 (microwave)	$N/H \leq 10^{-6}$ in 150 – 200 K range	less than solar	8, 9, 10, 11
CH_4 (1.1 μm, 1.27 μm, 1.5 – 2.5 μm, 6829 Å)	$\sim 20 \times$ solar[3]	up to $60 \times$ solar[12]	12, 13, 14, 15, 16
C_2H_6 (12.2 μm)	$< 3 \times 10^{-8}$ in the stratosphere[17]	3×10^{-6} in the stratosphere	17, 18
C_2H_2 (IUE, > 1600 Å) (Voyager, 1400 – 1600 Å)	$(6.8 \pm 3.4) \times 10^{16}$ cm^{-2} 2×10^{16} cm^{-2} at 0.3 mb		19 20
H (IUE, 1216 Å) (Voyager, 1216 Å)	$Ly\alpha = 1.4\,kR$[21] $Ly\alpha = 1.5\,kR$	$< 300\,R$[21]	21, 22, 23, 24 20

[1] Trafton (1976); [2] Hanel et al. (1986); [3] Tyler et al. (1986); [4] Macy and Smith (1978); [5] Trafton and Ramsay (1980); [6] Trafton (1978); [7] Lutz and Owen (1974); [8] Gulkis et al. (1978); [9] Gulkis et al. (1983); [10] Olsen and Gulkis (1978); [11] dePater and Massie (1985); [12] Pollack et al. (1986); [13] Encrenaz and Combes (1978); [14] Fink and Larson (1979); [15] Courtin et al. (1979); [16] Joyce et al. (1977); [17] Orton et al. (1983); [18] Macy and Sinton (1977); [19] Encrenaz et al. (1986); [20] Broadfoot et al. (1986); [21] Clarke et al. (1986); [22] Fricke and Darius (1982); [23] Clarke (1982); [24] Durrance and Moos (1982).

equilibrium products, so that their mixing ratios would vary with altitude. Only that C/H ratio in the atmospheres of Jupiter and Saturn (and most likely also in the Uranus and Neptune atmospheres) is greater than solar. If the atmospheric C/H ratio is representative of the bulk composition, its enhancement could imply an inhomogeneous accretionary process for the planetary formation. In-situ measurements of the numerous constituents of Jupiter's atmosphere in the mass range 1 – 150 AMU (atomic mass unit), to be carried out on Galileo/Jupiter Probe, will provide additional information critically needed for constructing meaningful models for the formation of the giant planets.

1.2 Outstanding Issues

Are the peculiarities of the "atmospheric" composition summarized in the last paragraph − C/H enrichment combined with possible O/H and N/H depletions on the major planets, He depletion on Saturn, presence of certain non-equilibrium species on Jupiter, but not on Saturn, etc. − indicative of the planetary bulk composition? If so, are the present theories of the origin and evolution of outer planet atmospheres adequate for explaining these and many other observed peculiarities? A satisfactory resolution of this issue would require deployment of deep entry probes at many locations on the planets to carry out in-situ composition measurements, including the isotopic ratios of the noble gases. High resolution spectroscopic observations with the Hubble Space Telescope can provide complementary information. Are there noteworthy seasonal, diurnal, and latitudinal variations in the bulk composition? Again, some suitably designed imaging spectroscopy observations at ultraviolet, visible, and infrared wavelengths from earth-orbit will be highly useful in this regard. Ground-based observations, particularly at the microwave and radio wavelengths, are an important complement to the data set.

A.1 Appendix

A.1.1 Escape and Thermal Velocities

An upward-moving particle escapes from a planetary atmosphere if its kinetic energy exceeds the planetary gravitational potential energy, i.e.,

$$\frac{1}{2}mv^2 \geq \frac{mMG}{R} ,$$

where m and v are, respectively, the mass and velocity of the particle, M is the planetary mass, G is the gravitational constant, and R is the radius at which escape occurs. The escape velocity, v_E is then obtained from the equality in the above expression, i.e.,

$$v_E = \sqrt{\frac{2MG}{R}} .$$

With,

$$g(R) = \frac{MG}{R^2} ,$$

where $g(R)$ is the planetary gravity at R,

$$v_E = \sqrt{2\,g(R)R} .$$

It is thus evident that the escape velocity is the same for all particles, irrespective of their masses. The mean thermal velocity, v_T, however, is dependent on the mass of the particle, since

$$\tfrac{1}{2} m v_T^2 = kT$$

so that,

$$v_T = \sqrt{\frac{2kT}{m}} ,$$

where k is the Boltzmann constant and T is the temperature.

The ratio of the gravitational energy to the thermal energy is referred to as the escape parameter, λ_{esc}, i.e.,

$$\lambda_{esc} = \frac{MmG}{RkT} = \frac{R}{H_e},$$

where H_e is the scale height of the escaping particle.

If thermal escape of a particle is to occur, its v_T should be comparable to or greater than v_E. Even for $v_T < v_E$, a certain fraction of the particles in the wings of v_T distribution may still be able to escape. This fraction is miniscule at Jupiter and Saturn, and amounts to less than 5% at Uranus and Neptune.

2 Thermal Structure

The thermal structure of the major planets is controlled mainly by the solar radiation, internal heat, and the auroral energy. In this chapter concepts of basic planetary temperatures and radiative transfer are developed first, followed by principles of the most commonly used techniques for temperature determination on the major planets. Since considerably more information is presently available for Jupiter and Saturn, their results are treated separately from Uranus and Neptune.

2.1 Equilibrium and Effective Temperatures, Internal Heat

The "equilibrium temperature" of a rotating planet is given by the Stefan-Boltzmann law:

$$T_{eq} = [F_i(1 - A)/4\,\sigma]^{1/4}\,, \tag{2.1}$$

where F_i is the solar flux incident on top of the planetary atmosphere, A is the Bond albedo (fraction of total incident radiation reflected by the planet), and σ is the Stefan-Boltzmann constant. For a planet which radiates energy at the rate it is received from the Sun, the "effective temperature" is equal to the equilibrium temperature. If the planet possesses an internal heat source, the effective temperature, T_{eff}, would be higher than the equilibrium temperature. In fact, the strength of internal energy is derived from measurement of the total planetary thermal emission integrated over all frequencies. The mean effective temperature is obtained using the following relationship

$$T_{eff} = (F_e/\sigma)^{1/4}\,, \tag{2.2}$$

where F_e is the planetary thermal emission flux.

The inner planets have virtually no internal heat compared to the total solar energy received by them. The situation with practically all of the major planets is just the opposite. The best estimates of the thermal emission flux at Jupiter and Saturn have been obtained from the data of Voyager Infrared Interferometric Spectrometer (Hanel et al. 1981b, 1983). Figure 2.1 shows a composite of the planetary and solar radiation behavior with wave number at Jupiter. The situation for Uranus and Neptune is less satisfactory due to the lack of phase and Bond albedo information. A consensus is, however, deve-

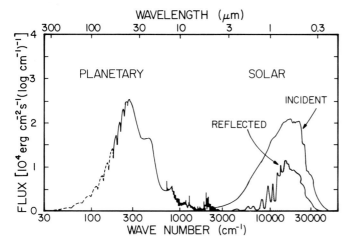

Fig. 2.1. Comparison of the Jovian IR emission flux with the incident and reflected solar spectrum at Jupiter. *Broken-line curve* below 230 cm^{-1} is a model fit to match the data between 200 and 300 cm^{-1}. (After Hanel et al. 1981b)

loping on the mean values of their Bond albedo, and the thermal emission from the sparse ground-based data in the infrared as summarized by Orton and Appleby (1984), Hildebrand et al. (1985) and Orton et al. (1986). For Uranus, some information is also provided by the Voyager IR measurements (Hanel et al. 1986). Table 2.1 lists our current understanding of the various quantities leading up to the derivation of the strength of internal heat on the major planets. For Jupiter and Saturn, the effective temperatures are far in excess of their equilibrium temperatures, implying that the total emitted energy is nearly twice the absorbed solar energy. For Neptune, this ratio is expected to be between 2 and 3. Uranus, on the other hand, does not seem to possess an internal heat source.

Gravitational contraction (called Helmholtz contraction) of Jupiter is responsible for its internal heat. Saturn – which is one third as massive as Jupiter – does not have adequate primordial heat to account for its internal energy. The excess is presumably supplied on release of the gravitational energy as liquid helium separates from metallic hydrogen in the interior of Saturn (Stevenson and Salpeter 1977; Stevenson 1980). Without such differentiation, Saturn would have decayed to its present luminosity in less than half the age of the solar system. It is curious why Neptune – which has virtually none of the internal energy sources operative on Jupiter and Saturn – still possesses a large internal heat reservoir. Tidal interaction between Triton and Neptune (Chap. 7 on Satellites) also cannot account for the internal energy on Neptune. It is even more intriguing that Uranus, whose size, mass, and atmospheric temperatures are expected to be similar to Neptune's, does not have an internal heat source.

Table 2.1. Internal energy

Quantity	Jupiter	Saturn	Uranus	Neptune
Incident solar energy, F_i [erg cm^{-2} s^{-1}]	5.076×10^4	1.506×10^4	3.72×10^3	1.52×10^3
Bond albedo, A	0.343 ± 0.032	0.342 ± 0.030	$0.31 - 0.35$	~ 0.33
Absorbed solar energy, $F_a = (1 - A)F_i$ [erg cm^{-2} s^{-1}]	3.35×10^4	9.99×10^3	$\sim 2.5 \times 10^3$	1.02×10^3
Total absorbed solar energy, $F_{at} = F_a \times A_c{}^a$ [erg s^{-1}]	5.014×10^{24}	1.114×10^{24b}	5.1×10^{22}	1.9×10^{22}
Equilibrium temperature, $T_{eq} = (F_a/4\,\sigma)^{1/4}$, [K]	110.2	82	~ 58	$46 - 50$
Emitted energy flux, F_e [erg cm^{-2} s^{-1}]	1.359×10^4	4.614×10^3	$\leq 7.1 \times 10^2$	$\sim 7.5 \times 10^2$
Total emitted energy, $F_{et} = F_e \times A_s{}^a$ [erg s^{-1}]	8.365×10^{24}	1.977×10^{24}	$\leq 5.8 \times 10^{22}$	$\sim 5.5 \times 10^{22}$
Effective temperature, $T_{eff} = (F_e/\sigma)^{1/4}$, [K]	124.4	95	≤ 59.4	60.3 ± 2
Internal energy flux $(F_{et} - F_{at})/A_s$ [erg cm^{-2} s^{-1}]	5.444×10^3	2.01×10^3	$\leq 7.2 \times 10^1$	4.8×10^2
Energy balance $(F_{et}/F_{at})^c$	$1.668\ (\pm 0.085)$	$1.78\ (\pm 0.09)$	≤ 1.14	$1.9 - 2.9$

[a] A_c = Cross-sectional area (disc); A_s = planetary area.
[b] Includes ring effects.
[c] Takes into account uncertainties in the Bond albedo. The value for Uranus is a preliminary estimate from Voyager observations.

2.2 Radiative Transfer

In the tenuous thermospheres of the major planets, heat conduction occurs down to an altitude where radiation can occur ($P \geq 10$ μb). The observed thermospheric temperatures (at least on Jupiter and Saturn), however, require energy input in addition to the solar EUV (see Sect. 2.4.2). The temperature structure in the troposphere and the middle atmosphere is controlled, by and large, by radiation equilibrium processes. In the deeper troposphere (around 0.5 to 1 bar) where optical thickness is large, thermal gradient becomes too large for the atmosphere to be stable, and convection begins to dominate. Still deeper in the troposphere, the dry adiabatic lapse rate is modified continuously as various volatiles begin to condense (see Chap. 3 on Cloud Structure).

The thermal opacity of He and H_2 is much greater than that of CH_4 in the troposphere, and it provides the dominant heat transport mechanism down to the pressure level of 0.5 to 1 bar. Weaker bands of CH_4, dust, and aerosols are also important in the tropospheric energy exchange. In the stratosphere

$(P \leq 100$ mb), radiative transport is still dominant. Local thermodynamic equilibrium (LTE) prevails because of a relatively large number of collisions. Heating occurs due to the absorption in the v_3 band of CH_4 at 3.3 μm, with subsequent re-radiation in the v_4 band at 7.8 μm. For better agreement with the observed emission intensities at 7.8 μm, most radiative equilibrium models require absorption also in the weaker bands of CH_4, and some type of dust or aerosol (Wallace 1976). The source of dust (or aerosol), sometimes referred to as Axel dust or Danielson dust in the context of Jupiter, is elusive. It could be photochemical haze formed on the condensation of hydrazine in the ammonia photochemistry (see Chap. 5 on Photochemistry), or it could be some type of extraplanetary material (e.g., from meteorites, rings, etc.) trapped in the vicinity of the inversion layer. On Jupiter and Saturn, cooling in the v_5 band of C_2H_2 at 13.7 μm, and the 12.2 μm band of C_2H_6 appears to be important near the homopause (Wallace 1976), and it is non-negligible below this level. The atmosphere does not remain in LTE at pressures lower than approximately 10 μb.

2.2.1 Radiative Transfer Equation

In the absence of conduction and convection, heat is transported by radiation; the atmosphere is said to be in radiative equilibrium when the incoming flux everywhere is balanced by the outgoing flux, i.e., the net flux divergence is zero. The rate of change of intensity I_v at frequency v along a slant path ds is given by

$$\frac{1}{\rho} \frac{dI_v}{ds} = -(k_v + \sigma_v)I_v + j_v, \tag{2.3}$$

where ρ is mass density of the homogeneous medium, $(k_v + \sigma_v)$ is called the extinction coefficient and it comprises the mass absorption coefficient, k_v, and mass scattering coefficient, σ_v; j_v is the emission coefficient [nomenclature of Chandrasekhar (1960) and Chamberlain (1978) is used for the most part]. In the absence of scattering ($\sigma_v = 0$) one can define a local thermodynamic equilibrium, so that the temperature everywhere is the same. The complete LTE approximation is valid only in regions where collisions dominate. For LTE conditions, the emission coefficient is related to the absorption coefficient by Kirchoff's Law, according to which

$$j_v = k_v B_v(T), \tag{2.4}$$

where $B_v(T)$ is the Planck's radiation function at temperature T and frequency v, it is given by

$$B_v(T) = \frac{2hv^3}{c^2} \frac{1}{\exp(hv/kT) - 1}, \quad \text{and} \quad \int B_v dv = \sigma T^4. \tag{2.5}$$

In this expression, h is the Planck's constant and c is the velocity of the light. The radiative transfer equation for LTE therefore takes up the following form for radiation in the (v, ϕ) direction

$$\frac{1}{k_v \rho} \frac{dI_v}{ds}(v, \phi) = -I_v(v, \phi) + B_v(T) . \tag{2.6}$$

For a plane-parallel atmosphere, the above equation can be expressed in terms of the vertical optical thickness, τ_v, using the following definition of τ_v

$$d\tau_v = -k_v \rho dz , \tag{2.7}$$

where

$$dz = ds/\cos v = ds/\mu , \tag{2.8}$$

or

$$d\tau_v = -k_v \rho ds/\mu ; \tag{2.9}$$

$\mu = \cos v$, where v is the zenith angle of the direction of radiation with the vertical (generally denoted as χ or ζ for solar zenith angle); z is measured vertically upward, and τ increases downward.

With the above definitions, the radiative transfer equation for a plane-parallel atmosphere in LTE is given as follows

$$\mu \frac{dI(\tau, \mu, \phi)}{d\tau_v} = I_v(\tau, \mu, \phi) - B_v(\tau, \mu, \phi) . \tag{2.10}$$

For a semi-infinite atmosphere [i.e., one which is bounded on one side and extends to infinity in the other direction] the emergent intensity is given by

$$I_v(0, \mu, \phi) = \int_0^\infty B_v(\tau, \mu, \phi) \exp(-\tau/\mu) \frac{d\tau}{\mu} . \tag{2.11}$$

The solution of the radiative transfer equation, however, is not as straightforward as above, since B_v itself is a function of the heating from radiation field I_v. For both monochromatic and "gray" atmospheres approximations can be made to "estimate" the radiation field (hence temperature) as a function of the atmospheric optical thickness, height or pressure. A gray atmosphere approximation is the one in which the absorption coefficient is independent of frequency — such flat, continuum behavior is most valid for applications in stellar atmospheres. For a plane-parallel atmosphere in LTE, the monochromatic radiative equilibrium equation can be solved by

  i. integrating Eq. (2.10) over a sphere, and
 ii. by multiplying Eq. (2.10) by μ, followed by integration over a sphere.

The first procedure yields

$$\frac{d}{d\tau_v}(F_v) = 4\pi(J_v - B_v),$$ (2.12)

where F_v represents the flux across a layer of the stratified atmosphere, i.e.,

$$F_v = (I^\uparrow - I^\downarrow)$$ (2.13)

(the up and down arrows correspond respectively to the outgoing and incoming components)

also,

$$F_v = 2\pi \int_{-1}^{+1} I_v(\mu)\,\mu\,d\mu.$$ (2.14)

J_v is the local mean intensity, i.e.,

$$J_v = \tfrac{1}{2}(I^\uparrow + I^\downarrow)$$ (2.15)

also,

$$J_v = \int I_v\,d\Omega,$$ (2.16)

where Ω is the solid angle.

The second procedure yields the following relationship between J_v and F_v

$$\frac{4}{3}\frac{dJ_v}{d\tau_v} = F_v,$$ (2.17)

substitution of which into Eq. (2.12) yields

$$\frac{d^2 F_v}{d\tau_v^2} - 3F_v = -4\frac{dB_v}{d\tau_v}.$$ (2.18)

For monochromatic radiative equilibrium, the first term in the above expression is zero, so that

$$\frac{dB_v}{d\tau_v} = \frac{3}{4}F_v = \text{const.}.$$ (2.19)

The same expressions hold true for the gray atmosphere approximation. Using Stefan-Boltzmann law and boundary conditions, one arrives at the following expression for the emission function at any optical thickness,

$$B_v(\tau) = B_v(T_0)(1 + \tfrac{3}{2}\tau),$$ (2.20)

where $B_v(T_0)$ is the value at $\tau = 0$; at this boundary, $T = T_0$. Using Eq. (2.20), Stefan-Boltzmann law [(Eq. 2.5)], and the fact that the total emergent flux from the atmosphere is twice that of a black-body at temperature T_0, which implies

$$T_{\text{eff}}^4 = 2 T_0^4,$$ (2.21)

it is found that

$$T^4(\tau) = \frac{T_{eff}^4}{4}(2 + 3\ \tau)\ .\tag{2.22}$$

Equation (2.22) is known as the Milne-Eddington equation.

The above procedure for estimating temperatures as a function of height (or optical depth) in the outer planets is very approximate. The actual thermal profile is obtained either by directly inverting the radiance data (as is routinely done for terrestrial problems; see, e.g., Goody 1964), or by modeling the equilibrium between the incoming radiation (composed of solar and the internal heats) and the outgoing radiation at each layer of the atmosphere (Trafton 1967).

The most successful models of the atmospheric thermal structure are the ones that use a hybrid approach in which results of the radiative convective equilibrium calculations are iterated with those obtained on direct inversion of the radiance data. This method has been used to retrieve thermal structure of the lower atmospheres of Jupiter and Saturn successfully.

2.3 Principle of Solar, Stellar, and Radio Occultations

The atmospheric and ionospheric height profiles, thermal structure, and the vertical mixing can be determined using the occultation technique. The planetary atmosphere serves as a long path length cell which can cause diminution or extinction of "light" passing through it. The source of light (electromagnetic radiation) is generally the Sun, some suitable star, or the spacecraft/satellite radio wave generator. At long wavelengths (such as visible and radio) the extinction results from differential refraction of the light by atmospheric molecules or electrons. At shorter wavelengths (e.g., ultraviolet), however, absorption (and some scattering) by the various atmospheric gases is mainly responsible for the extinction of light. By monitoring the diminution of light as it traverses successively deeper layers of the atmosphere, one can determine its structure. The atmospheric results obtained from this technique are independent of the knowledge of absolute flux of the source, as they depend only on the relative diminution of the light. On the other hand, some knowledge of the bulk composition of the atmosphere greatly facilitates the reduction and interpretation of the data. In the following sections the ultraviolet, visible, and radio occultations are treated individually in order to emphasize their applicability and differences.

2.3.1 Ultraviolet Solar and Stellar Occultations

A typical geometry of the UV occultation experiment is shown in Fig. 2.2. The detector mounted on the spacecraft/satellite measures the unattenuated inten-

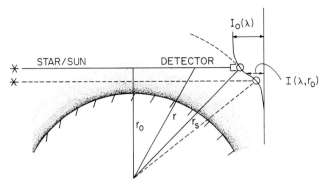

Fig. 2.2. Geometry of ultraviolet stellar and solar occultation experiments. The detector of the UV spectrometer aboard some satellite or spacecraft monitors the light from a suitable source, such as the Sun or a star. (After Hays et al. 1972)

sity, I_0, of the source when the line connecting the source and the detector, called minimum tangent ray, is above the atmosphere. As the spacecraft/ satellite moves in its trajectory, the minimum tangent ray descends deeper into the atmosphere, resulting in progressively greater absorption of light. The attenuation of monochromatic light of wavelength λ due to absorption by the ith gas of the atmosphere is given by Beer's Law, which states

$$I(\lambda, r) = I_0(\lambda) \exp[-\tau_i(\lambda)],\qquad(2.23)$$

where $I(\lambda, r)$ is the attenuated intensity at radius r, and $\tau_i(\lambda)$ is the line of sight optical depth of constituent, i, i.e.,

$$\tau_i(\lambda) = \sigma_i(\lambda) N_i(r).\qquad(2.24)$$

In the above expression $\sigma_i(\lambda)$ is the absorption cross-section of the ith constituent at wavelength λ, and $N_i(r)$ is its line of sight column abundance at radius r. For an isothermal atmosphere

$$N_i(r) = n_i(r) H \eta(r),\qquad(2.25)$$

where H is the atmospheric scale height, $n_i(r)$ is the number density of the ith constituent and $\eta(r)$ is the line of sight enhancement factor at radius r. $\eta(r)$ is given by

$$\eta(r) = \sqrt{2\pi r/H},\quad\text{(Hunten 1971)}.\qquad(2.26)$$

By carefully choosing the wavelengths, one can determine the atmospheric scale height in the occultation experiment. If λ_1 and λ_2 represent two wavelengths such that

$$\sigma(\lambda_1) \approx \sigma(\lambda_2),\qquad(2.27)$$

then equal amounts of attenuation would take place at only slightly different radii r_1 and r_2, i.e.,

$$\tau(\lambda_1) = \tau(\lambda_2) \,. \tag{2.28}$$

This also implies that

$$\eta(r_1) = \eta(r_2) \,. \tag{2.29}$$

Therefore, one obtains from Eq. (2.24) along with Eqs. (2.25) and (2.29)

$$\sigma_1 n_1 = \sigma_2 n_2 \,. \tag{2.30}$$

But,

$$n_2 = n_1 \exp(-\Delta z/H) \,, \tag{2.31}$$

where

$$\Delta z = z_2 - z_1 = r_2 - r_1 \,. \tag{2.32}$$

Hence,

$$H = \Delta z / \ln(\sigma_2/\sigma_1) \,, \quad \text{(Atreya et al. 1979a)} \,. \tag{2.33}$$

Knowing the scale height one can determine the temperature of the isothermal region.

The above analysis, however, is extremely restrictive, since the conditions of isothermal regime, nearly equal absorption cross-sections, single absorber, etc. cannot always be met. In general, therefore, atmospheric information from the stellar occultation light curves is obtained by proceeding from Eq. (2.24) and inverting the line of sight column abundance, N. For a spherically stratified atmosphere,

$$N_i(r_0) = 2 \int_{r_0}^{\infty} \frac{n_i(r) \cdot r\,dr}{(r^2 - r_0^2)^{1/2}} \,, \quad \text{(Hays and Roble 1968)} \,. \tag{2.34}$$

Inversion of the above line of sight column abundance can be done by using the Abel inversion method which gives

$$n_i(r) = \frac{d}{dr} \left(-\frac{1}{\pi} \int_{r}^{\infty} \frac{r}{r_0} \frac{N_i(r_0)\,dr}{(r_0^2 - r^2)^{1/2}} \right) \,. \tag{2.35}$$

For most atmospheric applications, the numerical iterative method for retrieving densities is found to be more suitable than the analytical inversion technique (Atreya et al. 1976; Atreya 1981).

The source of ultraviolet radiation in the occultation experiment can be either the Sun or a suitable star. The Sun has the advantage of having large continuum flux at most wavelengths. Its large angular dimension at planetary distances can, however, have the consequence of averaging atmospheric information over several scale heights, as is evident from the actual occultation geometry presented in Table 2.2 for recent solar occultation experiments at the major planets. The solar occultation experiment appears most promising only for Uranus, Titan, and Triton (Table 2.2). If only a small portion of the solar disc is used, it is important to ensure that the observed attenuation is

Table 2.2. Characteristics of the best solar occultation experiments from Voyager

Planet	Solar diameter at the planet [deg]	Spacecraft range to point of tangency [km]	Projected size of the Sun in the atmosphere [km]	Average exospheric scale height [km]
Jupiter	0.102	4.9×10^5	880	~200
Saturn	0.056	1.5×10^5	150	175 – 350
Titan	0.056	3360	3.3	~100
Uranus	0.028	2.4×10^5	120	500
Neptune	0.018	$\sim 5 \times 10^5$	160	≥ 50 (?)
Triton	0.018	5000	1.5	?

actually due to the atmospheric absorption, and not because a different part of the Sun's disc was in the field of view during occultation. This is due to the fact that the UV brightness distribution is not uniform over the solar disc. Stellar occultations, on the other hand, do not suffer from the above disadvantages of the solar occultations, as the stars are merely point sources. The choice of suitable stars, i.e., those with relatively constant and large continuum flux in the ultraviolet, is, however, rather restricted. Moreover, because of the interstellar hydrogen absorption, the available flux is limited to wavelengths longer than 911 Å, the Lyman cut-off limit.

2.3.2 Visible Stellar Occultation

Unlike UV stellar occultations, here the differential refraction of visible starlight by density gradient in the atmosphere results in the extinction of the signal. Occultation in the visible can be monitored from ground, providing information on turbulence and thermal structure of planetary atmospheres at pressures generally greater than a microbar. A typical occultation geometry is shown in Fig. 2.3. The tangent rays which pass closer to the planet undergo greater refraction due to the greater atmospheric density than those farther from the planet. In an isothermal atmosphere with scale height H, the angle of refraction for a ray passing at planetocentric distance r_1 is given by

$$\theta = v(r_1) \, [2 \, \pi r_P / H]^{1/2} , \qquad (2.36)$$

where r_P is the planetary radius, taken generally as the radius at the surface, cloud tops or the 1-bar level; $v(r_1)$ is the atmospheric refractivity at r_1, also defined as

$$v(r_1) = \mu(r_1) - 1 , \qquad (2.37)$$

where $\mu(r_1)$ is the refractive index. Since μ is close to unity for most planetary gases, v (rather than μ) is used to emphasize effects of small differences in μ. The values of refractivity for an ($H_2 +$ He) mixture (outer planets) is

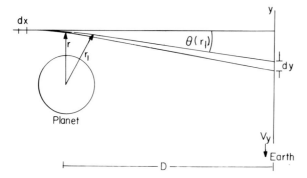

Fig. 2.3. Geometry of a typical stellar occultation in the visible monitored from the ground. The refractive bending, $\theta(r_1)$, of a ray passing at a distance r_1 is smaller than that of a ray passing closer to the planet. At 1 AU, the two rays diverge by a distance dy. dx is an increment of pathlength along the direction of propagation. The planet-Earth system is considered at rest, whereas the Earth moves with a velocity of V_y in the y-direction. (After Elliot et al. 1974, and Hunten and Veverka 1976)

$\sim 1.2 \times 10^{-4}$, and $\sim 4.5 \times 10^{-4}$ for CO_2 (Mars, Venus) at Standard Temperature and Pressure (STP). The refractivity is proportional to the atmospheric density, so that

$$v(r_1)/n(r_1) = v(r_P)/n(r_P) .\qquad (2.38)$$

Once the bending angle is known, one can determine $v(r_1)$ and hence $n(r_1)$. The most significant atmospheric information is obtained below what is known as the occultation level, or the half-intensity level. This level occurs where

$$\theta = H/D \quad \text{(Hunten and Veverka 1976)} ,\qquad (2.39)$$

where D is the observer's distance from the planet. For a stellar occultation by Jupiter ($T = 170$ K in the middle atmosphere, $H \approx 30$ km), monitored from the ground, ($D = 5.2$ AU), the bending angle, θ, is typically on the order of 4×10^{-8} radians. This implies that $v(r_1) = 3 \times 10^{-10}$ [using Eq. (2.36)], and $n(r_1) = 7 \times 10^{13}$ cm^{-3}, or approximately the 2 µb level. Thus the best data for Jupiter are obtained at pressures generally greater than 1 µb, (Elliot et al. 1974). An excellent discussion of the inversion techniques and error analysis of the stellar occultation data has been published by Hunten and Veverka (1976).

2.3.3 Radio Occultation

In radio occultations, generally the S (13 cm-λ, or 2.293 GHz) and X (3.5 cm-λ, or 8.6 GHz) bands of the spacecraft telemetry serve as the source of "light". Since these occultations occur close to the planet (i.e., D is small), the bending angles are generally large – typically 10^{-4}, compared to 10^{-8} radians

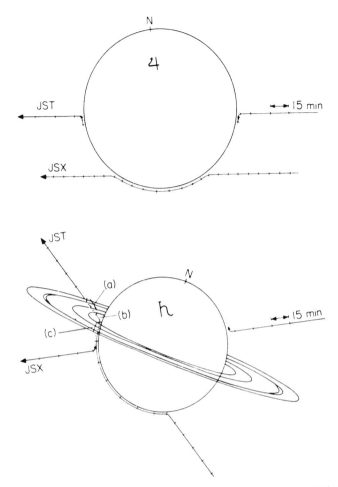

Fig. 2.4. *Upper diagram* View from Earth of Voyager radio occultations. The JST (Jupiter, Saturn, Titan) trajectory provides a more central, while the JSX (Jupiter, Saturn, Uranus) trajectory results in a more grazing occultation. *Lower diagram* side view of Voyager occultations at Saturn. Region (*a*) provides a clear ring occultation, region (*b*) a clear atmospheric occultation, and region (*c*) a combination of the two. (Eshleman et al. 1977)

for the visible stellar occultations. Thus occultation level is correspondingly deeper in the atmosphere, i.e., around the millibar level. Another way of looking at it is to determine the occultation level based on the measurement limit for radio refractivity, which is $\sim 10^{-6}$. It should be noted that in radio applications, refractivity is defined as $(\mu - 1) \times 10^6$. Moreover,

$$\mu - 1 = n_m \bar{v}_m , \tag{2.40}$$

where n_m and $\bar{v}_m$ are respectively the number density and the refractive volume of the molecules. Typically, $\bar{v}_m = (0.5 - 5) \times 10^{-23} \, \text{cm}^3$, so that $(\mu - 1) \approx 10^{-6}$

would occur around the $10^{17}\,\text{cm}^{-3}$ density level, or ~ 1 mbar level on Jupiter. The effects of absorption and multipath propagation limit the pressure upper limit for detection to around 1 bar at Jupiter and Saturn and 3 to 5 bar at Uranus and Neptune (Eshleman 1984, personal communication). In the ionosphere, refraction of the radio waves is caused by electrons which have a considerably smaller refractive volume, $\bar{v}_e$, of only $10^{-12}\,\text{cm}^{-3}$, ($\bar{v}_e = r_e \lambda^2 / 2\,\pi$, where r_e is the classical radius of electrons $= 2.8 \times 10^{-13}$ cm; and the radio wavelength, $\lambda = 3.6$ or 13 cm). Hence, even extremely low values of the electron concentrations (100 to 1000 cm^{-3}) can be detected in the ionosphere.

The measured refractivity in the neutral atmosphere can be related to the ratio of the mean molecular weight to the temperature ($\bar{m}/T$). If the temperature or the composition is constrained at some altitude, then information on the other quantity can be obtained from the radio occultation data. For the neutral atmosphere, the boundary values for analysis of the data are generally provided by simultaneous measurements done in the infrared.

A typical geometry for the Voyager radio occultations at the outer planets is shown in Fig. 2.4. It applies to refractions in both the neutral and the ionospheric regions. The actual measurement parameters in radio occultation experiments are Doppler frequency shift, group delay, intensity, and polarization of the radio signal.

2.4 Temperature Measurements – Jupiter and Saturn

The most extensively studied atmospheric regions are the troposphere above the ammonia clouds, and the lower stratosphere. The information for the upper atmosphere is limited to that available from a few successful occultation experiments. The results for the two atmospheric regions are discussed below.

2.4.1 Lower Atmosphere

The infrared observations done from Pioneer as well as Voyager spacecraft have yielded the thermal structure of the lower stratosphere and the troposphere (Ingersoll et al. 1976; Orton and Ingersoll 1976; Hanel et al. 1979a, b, 1981a, 1982). Additional information was obtained from the Voyager radio refractivity data (Eshleman et al. 1979a, b; Tyler et al. 1981, 1982; Lindal et al. 1981). The Voyager infrared data are by far the most extensive.

The thermal radiance data most useful in deriving the tropospheric temperatures are the pressure-induced $S(0)$ and $S(1)$ translation-rotation transitions of H_2 between 300 and 700 cm^{-1}. This is because H_2 alone contributes to the opacity in this region. The ν_4 band of CH_4 at 1306 cm^{-1} gives additional information, primarily in the lower stratosphere. For Saturn, however, low signal-to-noise ratio, caused by the low stratospheric temperature, necessitated averaging of several tens of spectra for retrieving thermal information

Table 2.3. Mean atmospheric temperatures [K]

Region	Jupiter	Saturn	Uranus	Neptune
Troposphere, 1 bar	165 ± 5[a]	134	76 – 78	75 – 78
Tropopause	110 (at ~140 mb)	85 (at ~100 mb)	52 – 54 (at ~100 mb)	53 – 55 (at ~200 mb)
Lapse Rate[b] [K km^{-1}]	1.9	0.8	0.7	0.85
Stratosphere	160 ± 20 (at 10 – 1 mb)	140 – 150 (at 1 mb)	88 – 90 (at 1 mb)	140 – 150 (at 0.1 mb, model)
Upper Atmosphere				
Mesopause (?)	200 ± 50 (at ~1 μb)	140 ± 20 (at 0.01 μb)	120 – 155 (at ~1 μb)	140 – 150 (at ~1 μb)
Exosphere	1100 ± 200	800 ± 150[c]	750 ± 100	≥ 140

[a] Adiabatic extrapolation of Voyager data (Lindal et al. 1981). Hunten et al. (1980) derive 156 – 159 K.
[b] Average dry adiabatic lapse rate, Γ_d, in the troposphere at midlatitudes, assuming solar composition (see Chap. 3 for additional details).
[c] Stellar occultation results (Festou and Atreya 1982). Smith et al. (1983) give 420 ± 30 K from solar occultation, while radio occultation plasma scale heights imply 600 – 1000 K.

from the v_4 band of CH_4 (Hanel et al. 1981a, 1982). The lapse rate is found to be close to adiabatic in the region below ~300 mb on Jupiter, and ~500 mb on Saturn. Indeed, transport of internal heat to the visible atmosphere necessitates the lower troposphere to be convective and the tropospheric lapse rate adiabatic (Stone 1973). The globally averaged temperature structure in the main atmospheric regions of the major planets is summarized in Table 2.3. Important latitudinal, seasonal and other variations are discussed below.

Jupiter

Hanel et al. (1979a, b) have concluded from the analysis of their Voyager infrared data that large spatial variability exists in the tropospheric/stratospheric thermal structure of Jupiter. A typical thermal structure for two equatorial latitudes and the Great Red Spot (GRS is discussed in Appendix A.2.1) is shown in Fig. 2.5. In addition, vertical structure was derived from zonally averaged meridional cross-sections by using north-south scans. Horizontal structure was obtained from measurements of global brightness temperatures at 602 cm^{-1} (corresponding to ~800 mb level) and 226 cm^{-1} (corresponding to ~150 mb level). The important features of the thermal structure are discussed below (Hanel et al. 1979a, b, 1981b).

  i. The above data, along with the imaging and radio observations, indicate that the NH_3 cloud tops are located at around the 600 mb level, where the temperature is approximately 150 K.

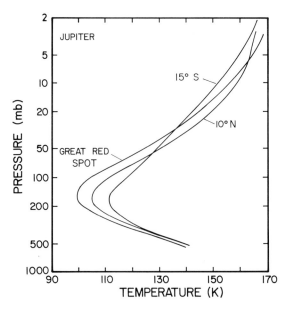

Fig. 2.5. Jovian tropospheric and lower stratospheric thermal profiles retrieved from the Voyager IRIS data. (Hanel et al. 1979a; copyright 1979 by the AAAS)

ii. The tropopause temperature decreases, and the height of the tropopause increases roughly from the equator to the pole. The equator-to-pole cooling is expected for a warmer lower stratosphere overlying a relatively stable troposphere. The warmest regions in the troposphere are found to be between 15°N and 15°S, a region which is associated with the equatorial belt (Appendix A2.2). The coolest regions lie between 20°N and 35°N, and between 20°S and 35°S, both of which are associated with the regions of zones (Appendix A2.2).

iii. The upper stratosphere is found to be cooler at high southern than the high northern latitudes, indicating a pole-to-pole meridional circulation and radiative control by seasonal insolation.

iv. The upper troposphere and the lower stratosphere above the GRS are colder than in the surrounding regions. The maximum horizontal thermal contrast at the tropopause is 5 to 7 K.

Finally, a comparison between the Voyager 1 radio occultation and the infrared-deduced temperatures is shown in Fig. 2.6 (Lindal et al. 1981). Although there is general agreement between the results obtained from the two techniques, there are notable differences. The radio data clearly indicate warm inversion layers around the 35 mb (ingress) and 3 to 10 mb levels. Such inversions are most likely the result of absorption of the solar radiation by dust or aerosols layers.

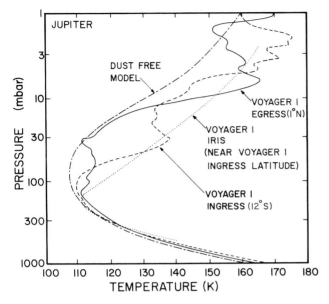

Fig. 2.6. Comparison of the lower atmospheric temperatures at Jupiter obtained from Voyager infrared and radio experiments. The Voyager 1 ingress radio data correspond to late afternoon near the South Equatorial Belt, whereas the egress data are for the predawn period near the Central Equatorial Current region. (Lindal et al. 1981)

Saturn

The average tropospheric and stratospheric thermal structures deduced from the Voyager infrared (IRIS) and radio (RSS) investigations are generally in agreement, as is evident from Fig. 2.7. There are, however, some distinct differences, e.g., comparison of the IRIS 13°N and 22°N data with the RSS 31°S and 36°N data reveals that the IRIS temperatures are cooler than the RSS temperatures at ~700 mb, but warmer than the RSS temperatures in the upper atmosphere (100 – 10 mb region). Presence of hazes, aerosols, and ammonia clouds can be responsible for the underestimation of the lower tropospheric temperatures (Conrath and Pirraglia 1983). However, the apparent warming in the upper atmosphere is puzzling, since West et al. (1983) have found, using Voyager and Pioneer photopolarimeter data, little or no evidence of aerosols at mid-southern latitudes. If they were present, they could absorb the solar radiation, resulting in warming.

The latitudinal behavior of temperatures (Fig. 2.8) in the deep (~730 mb), middle (~290 mb) and the upper (~150 mb) troposphere has been obtained from radiative transfer inversion of the infrared opacities in the 200 – 600 cm^{-1} region (Hanel et al. 1981a, 1982). With the exception of mid-northern latitudes which are ~5 K warmer, there is virtually no equator-pole thermal gradient in the deep troposphere (~730 mb). Hazes or partial clouds may

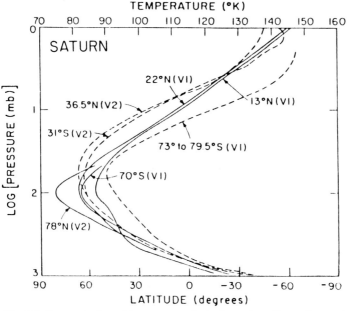

Fig. 2.7. Tropospheric and lower stratospheric thermal profiles at Saturn. *Solid lines* represent Voyager 1 (V1) and Voyager 2 (V2) infrared data (Hanel et al. 1981a, 1982); *dashed lines* are from the V1 and V2 radio data (Tyler et al. 1981 and 1982). (Prinn et al. 1984)

be responsible for the needed additional opacity in the mid-northern latitudes (Ingersoll et al. 1984).

The data for the 290 mb and 150 mb levels indicate presence of small-scale meridional gradients in temperature which are symmetric about the equator (Fig. 2.8). These small-scale structures are most likely actual thermal variations along constant pressure surfaces. They can result from some dynamical process such as zonal jets at the lower boundary below the upper troposphere (Ingersoll et al. 1984). The small-scale structure in the temperature is superimposed on a large-scale hemispherical asymmetry which indicates a definite equator-pole cooling in the northern hemisphere, while no such cooling is evident in the south. Furthermore, southern latitudes poleward of 25°, particularly for the upper troposphere ($\sim$150 mb), are 6 to 10 K warmer than the corresponding latitudes in the north. Such asymmetry results from lagged response to the solar forcing which has a seasonal variation on Saturn. [Because of its 27° orbital inclination, Saturn, like the Earth (orbital inclination, 23°), has four seasons. The Sun crossed Saturn's equator moving northward 8 months prior to Voyager 1 encounter and 17 months prior to Voyager 2 encounter. Thus, Saturn was entering its 7-year phase of spring at the time of Voyager encounters.] Conrath and Pirraglia (1983) have examined the seasonal response in terms of the radiative relaxation time, τ_R. The phase lag, δ, of the thermal response to solar forcing is given by

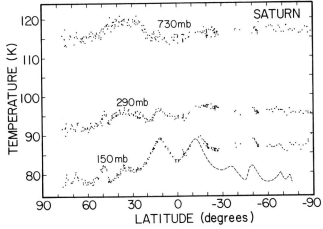

Fig. 2.8. Latitudinal behavior of the temperatures at three different pressure levels in the troposphere of Saturn. *Dashed line* shows a fit to the temperature in the northern hemisphere at 150 mb folded over at the equator for comparison with the southern hemisphere temperatures at corresponding latitudes. (After Conrath and Pirraglia 1983)

$$\delta = \tan^{-1}(\omega_s \tau_R) , \tag{2.41}$$

where ω_s is the seasonal frequency. The amplitude of response is proportional to $(\omega_s \tau_R)^{-1}$ for large values of $\omega_s \tau_R$. For the upper troposphere of Saturn (~ 150 mb), the thermal response is calculated to be approximately 5 years (Conrath and Pirraglia 1983; Gierasch and Goody 1969). Thus, the 150 mb temperatures in the northern hemisphere are expected to be about 8 to 10% [or by a factor of $(\cos 54°)^{1/4}$] cooler than in the south (Ingersoll et al. 1984). This is found to be consistent with the observations (Fig. 2.8). At higher pressures, the $\omega_s \tau_R$ term becomes large and the seasonal response begins to vanish. For example, at the 730 mb level (Fig. 2.8), the hemispheric difference is nonexistent, except for the mid-northern latitude warming discussed above.

In summary, the seasonal characteristics observed in Saturn's temperature structure are distinctly different from those in Jupiter's temperature structure, because the obliquity of the latter is virtually nil.

2.4.2 Upper Atmosphere

Ground-based stellar occultations in the visible, as well as Voyager/ultraviolet stellar and solar occultations have provided useful data for the thermal structure of the upper atmosphere. The most widely studied visible stellar occultation is that of star β Scorpii by Jupiter. A typical light curve for the emersion geometry is shown in Fig. 2.9. Temperature information between 0.1 μb and 30 μb levels (corresponding densities 10^{13} to 10^{15} cm^{-3}) deduced from the β

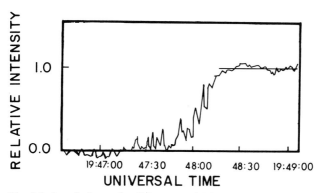

Fig. 2.9. A typical emersion light curve of β Sco occultation by Jupiter at −58° latitude. (After Veverka et al. 1974)

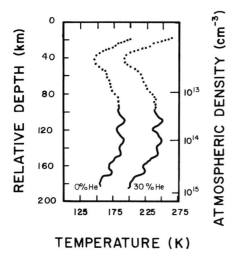

Fig. 2.10. Middle atmospheric temperature profiles at Jupiter retrieved from the β Sco occultation data. Curves are shown for 0% and 30% mole fractions of helium. (After Veverka et al. 1974)

Sco occultation light curves is shown in Fig. 2.10. There are two remarkable points about the retrieved temperatures: first, these are the only set of data available for the upper stratosphere/lower mesosphere of Jupiter; and second, the oscillations in the thermal profile (Fig. 2.10) are indicative of vertically propagating inertia gravity waves (French and Gierasch 1974). The latter is important for understanding the energy budget of the upper atmosphere. Since the He/H$_2$ ratio at Jupiter has now been measured to be 13.6%, it would seem from Fig. 2.10 that the average temperature is between 175 and 200 K in the microbar region. This is in good agreement with the Voyager results discussed next.

Occultations in the ultraviolet of several stars by the outer planets have been monitored by the Ultraviolet Spectrometer (UVS) on board Voyager

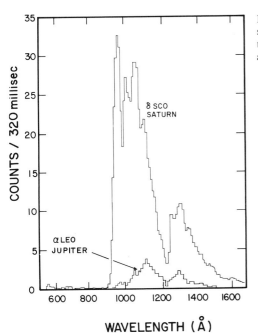

Fig. 2.11. The spectra of δ Sco and α Leo stars as recorded by Voyager 2 UV spectrometer. δ Sco was occulted by Saturn and α Leo by Jupiter

spacecrafts (Broadfoot et al. 1979, 1981a; Sandel et al. 1979, 1982a). The most successful of these are the equatorial occultations of star Regulus (α Leo, B7 type) by Jupiter on 9 July, 1979, and of star Dzuba (δ Sco, B0 type) by Saturn on 25 August, 1981. The unattenuated spectra of these stars as recorded by UVS are shown in Fig. 2.11. The UV flux of δ Sco is ten times greater than that of α Leo, resulting in better statistics of the Saturn data. The sudden extinction of signal below the Lyman cut-off limit (911 Å) means that the wavelength region of H_2 continuum absorption which could be employed in the solar occultation experiment is not available here. Instead, band absorption features are used to derive the H_2 densities. The decrease in the signal above 1200 Å in Fig. 2.11 is in response to the decreasing instrument sensitivity at these wavelengths.

The spatial and spectral resolution achieved in the above occultation experiments are interdependent. Although the projected size of the star in the atmospheres of Jupiter and Saturn is less than 100 m, it is the rate of descent of the minimum tangent ray in the atmosphere, combined with spectral sampling rate during occultation, that determines the height resolution. For both Jupiter and Saturn, the former is ~ 10 km s^{-1}, while the latter, 320 ms, providing a height resolution of 3.2 km, which is much smaller than the scale heights. The UVS is an objective grating spectrometer with 128 reticon-vidicon detectors covering a range of 512 to 1690 Å, i.e., each detector images approximately 9.2 Å. The spectral resolution is found to be 33 Å for a spatially ex-

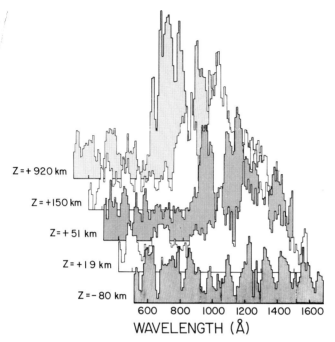

Z = + 920 km

Z = +150 km

Z = + 51 km

Z = + 19 km

Z = - 80 km

600   800   1000   1200   1400   1600

WAVELENGTH (Å)

Fig. 2.12. Solar occultation at Jupiter. Note the progressively greater absorption of the solar flux as the minimum tangent ray descends in the atmosphere. Altitudes are arbitrary and correspond to some convenient experimental scale. (After Festou et al. 1981)

tended source filling the UVS field of view, while it is 25 Å for a monochromatic point source. The entire UVS spectral region from 512 to 1690 Å is imaged every 320 ms during the above occultation experiments.

An examination of the solar spectrum during occultation, shown in Fig. 2.12, gives a fairly good indication of progressively greater absorption as the tangent ray descends deeper and deeper in the atmosphere of Jupiter. At 920 km above an arbitrary experimental altitude the solar flux is virtually unattenuated. At − 80 km, nearly all of the flux has been absorbed by the atmospheric molecular species. Atmospheric information is obtained by examining individual light curves, or a set of them added together to improve statistics. Figure 2.13, for instance, shows two distinctly different light curves obtained from the α Leo occultation experiment. Figure 2.13a for 1282 − 1393 Å is obtained by adding 12 channels, and Fig. 2.13b (1050 − 1152 Å) by adding 11 channels. The apparent rise in the unattenuated signal prior to the occultation in Fig. 2.13 is the result of changing background level which is mainly composed of electron-induced noise, interplanetary background, and emission from the Jovian disc. The relatively rapid extinction of signal in Fig. 2.13a is the result of absorption by starlight by any combination of the hydrocarbons CH_4, C_2H_6, C_2H_2 and C_2H_4 whose abundances increase rapidly below a

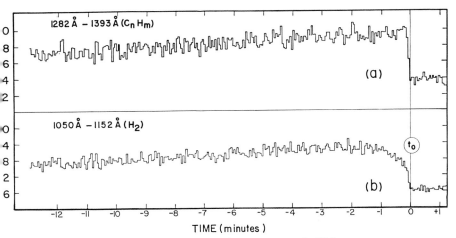

Fig. 2.13. α Leo occultation light curves at Jupiter. (Festou et al. 1981)

certain altitude called the homopause, (Chap. 4 on Vertical Mixing). The relatively gradual decrease of the signal in Fig. 2.13b is indicative of a constituent which absorbs to high altitudes in the atmosphere – it turns out to be H_2.

The H_2 density and temperature are determined by analyzing light curves in the $912-1200$ Å region. The absorption in this region is almost entirely due to the H_2 Lyman and Werner bands which connect the ground state $X^1\Sigma_g^+$ with the excited states $B^1\Sigma_g^+$ and $C^1\pi_u$. The observed transmission characteristics are simulated by calculating the transmission through an H_2 atmosphere with varying combinations of line of sight column abundances and temperatures. The transmitted intensity at frequency v is related to the unattenuated intensity as follows

$$I(v) = I_0(v) \exp\left[-\sum_{J'J''v'v''} k(v)l\right],$$ (2.42)

where l is the line of sight "depth" of the gas in cm, J, and v are respectively rotational and vibrational quantum states, and $k(v)$ is the absorption coefficient in cm^{-1}. Considering each individual rotational-vibrational line of the H_2 band system as having a Voigt profile (natural and Doppler broadenings only), the absorption coefficient can be defined as

$$k(v) = k_0 K(x, y),$$ (2.43)

where K is the Voigt function, and

$$x = \frac{v - v_0}{\alpha_D} \; (\ln 2)^{1/2}$$ (2.44)

$$y = \frac{\alpha_L}{\alpha_D} \; (\ln 2)^{1/2}.$$ (2.45)

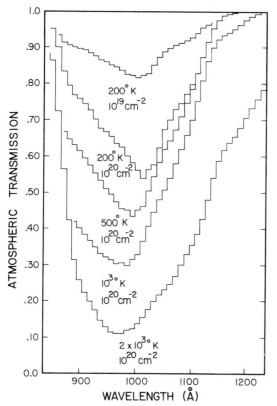

Fig. 2.14. Simulation of atmospheric transmission for Voyager 2
UV spectrometer in the wavelength
region where H_2 absorbs in its
Lyman and Werner bands. The
curves are for various combinations of (isothermal) temperatures
and column abundances in the line
of sight. (Festou et al. 1981)

v_0 is the line center wave number for a given fine structure transition and α_L and α_D are the Lorentz and Doppler line widths (full widths at half maximum intensity or FWHM, as discussed in Appendix A2.3). The quantity k_0 for a given line is related to the integrated line strength $S^{J''J'}$ as follows

$$S^{J''J'} = \alpha_D \left(\frac{\pi}{\ln 2}\right)^{1/2} \quad k_0 = h v_0 B_{J''J'} N_{J''}, \tag{2.46}$$

where h is the Planck's constant, $B_{J''J'}$ is the Einstein absorption probability for a given $J''J'$ transition of a particular band, and $N_{J''}$ is the population density of the transmission function in H_2. The H_2-band transmission characteristics achieved in the above manner are shown in Fig. 2.14. It is found that in the 150 – 2000 K range, the maximum absorption occurs around 960 Å. Hence, from a statistical viewpoint the wavelength region 912 – 1004 Å is the most appropriate to use for determining H_2 temperatures and densities. In the actual stimulation for comparison with the data, however, the temperature was allowed to vary along the line of sight, and additional wavelength regions (1041 – 1106 Å, and 1106 – 1189 Å) were used. This analysis yields H_2 density

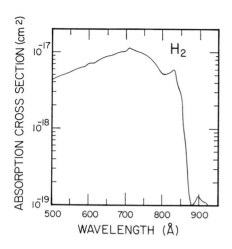

Fig. 2.15. H_2 continuum absorption cross-sections

and temperature in the 330 – 830 km above the NH_3-cloud tops (located at 600 mb, $T = 150$ K level). The H_2 density at 380 km is found to be 3(+4, −1) $\times 10^{13}$ cm^{-3} with corresponding temperature of 200 ± 50 K. The exospheric temperature, determined in a solar occultation experiment, is found to be 1100 ± 200 K (Atreya et al. 1981). To reconcile the thermospheric temperature and H_2-density results of the stellar occultation with the above exospheric temperature, a temperature gradient of approximately 1 K km^{-1} is required above about 400 km altitude. The exospheric temperature was derived from the solar occultation experiment using the light curves between approximately 600 and 700 Å, since the H_2 absorption cross-section does not vary greatly in this region (Fig. 2.15). The results, however, had large uncertainties (± 20%) due to the difficulties associated with solar occultation experiments, as discussed earlier.

The analysis of the hydrocarbon absorptions is complicated due to the fact that CH_4, C_2H_6, C_2H_2, and C_2H_4 all have nearly equal and large absorption cross-sections up to about 1400 Å (Fig. 2.16). Beyond 1400 Å, peculiar features in the C_2H_2 absorption cross-section can be used, in principle, to derive the C_2H_2 densities. It is, however, statistically difficult due to the very low sensitivity of the UVS at these long wavelengths. Another feature in the C_2H_2 absorption cross-section which has been successfully exploited is between 1250 and 1350 Å. In fact, a careful simulation of the transmission characteristics in small wavelength intervals (1265 – 1320 Å, 1310 – 1360 Å, 1365 – 1410 Å, and 1430 – 1480 Å) has permitted determination of the hydrocarbon abundances over a 50 km range at around the microbar level. The volume mixing ratios are found to be, CH_4: 2.5 (+3, −2) $\times 10^{-5}$ at 325 km, C_2H_6: 2.5 (+2, −1.5) $\times 10^{-6}$ at 325 km, and $C_2H_2 < 2.5 \times 10^{-6}$ at 300 km (Festou et al. 1981; Atreya et al. 1981). The Jovian temperature and density results obtained from the α Leo stellar occultation, along with the important data from the infrared and radio science experiments, are summarized in the "model atmosphere"

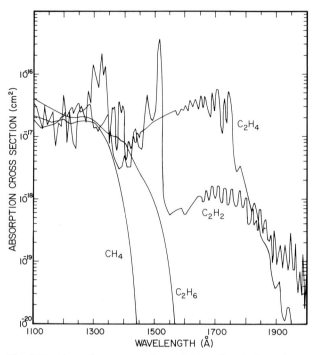

Fig. 2.16. Absorption cross-sections of the various hydrocarbon molecules of relevance to the major planets

shown in Fig. 2.17. The broken lines are interpolations between the lower and the upper atmospheric data. Similar analysis has been carried out for the data obtained from the occultation of star δ Sco by Saturn (Festou and Atreya 1982). Larger-scale heights in the mixed atmosphere, however, have not yet permitted deduction of all of the hydrocarbon densities. The CH_4 volume mixing ratio is found to be 1.5×10^{-4} at 965 km above the 1-bar level, and analysis of the same data yields the exospheric temperatures of 800 ($+150$, -120) K at 1540 km. Additional work is needed to resolve an apparent discrepancy between the above-mentioned equatorial exospheric temperature of Saturn derived from the stellar occultation data (Festou and Atreya 1982), and a lower value of 420 ± 30 K derived by Smith et al. (1983) from the solar occultation data at 30°N. The topside electron temperatures using the ionospheric radio occultation data are found to be 565 K at 36.5°N and 617 K at 31°S (Tyler et al. 1982). The present model atmosphere of Saturn for equator to midlatitude regions is presented in Fig. 2.18.

The middle atmospheric region — from 1 mb to 1 μb at Jupiter and 1 mb to 10 nb (nanobar) at Saturn — has not yet been probed. It is imperative that future spacecraft missions, such as Cassini Saturn Orbiter/Titan Probe, include instrumentation capable of detecting extremely weak absorptions (e.g.,

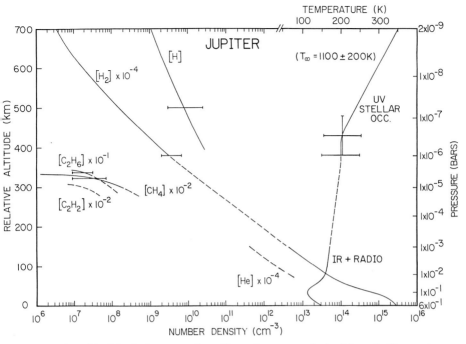

Fig. 2.17. A model of the Jupiter atmosphere. Thermal structures obtained from the Voyager infrared, radio, and ultraviolet data are shown. *Broken line* interpolations in temperature and density represent the region of information gap. The hydrocarbon data in the upper atmosphere are from the Voyager ultraviolet stellar occultation. (Atreya et al. 1981)

in the wings of the hydrocarbon absorption bands) and airglow emission features (e.g., resonance lines of any trace species like Si, SiO, metallic ions, etc. that may be present in the atmosphere). This would necessitate moving to longer ultraviolet wavelengths, up to 2000 or 3000 Å, for occultation experiments, and building up airglow statistics with repeated observations from an orbiter – some such observations are anticipated with Galileo/Ultraviolet Spectrometer.

2.5 Temperature Measurements – Uranus and Neptune

Substantial information on the atmospheric temperature structure of Uranus and Neptune has been obtained from ground-based observations of infrared thermal radiance (for $P \gtrsim 0.1$ mb) and the visible stellar occultations (for $P \gtrsim 1$ µb). An example of a typical stellar occultation by Neptune is shown in Fig. 2.19. The Voyager ultraviolet solar and stellar occultations have provided valuable data for the upper atmosphere (Broadfoot et al. 1986), whereas the radio occultation and the infrared experiments have yielded the thermal pro-

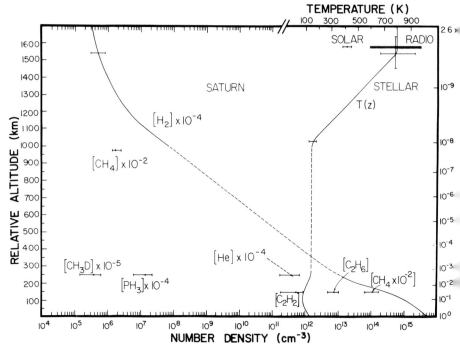

Fig. 2.18. Same as Fig. 2.17, except for Saturn. Some principal minor species detected in the infrared in the lower atmosphere are also indicated. (After Festou and Atreya 1982; Atreya et al. 1984)

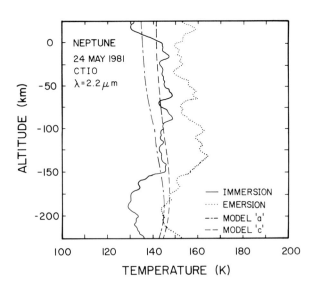

Fig. 2.19. Temperatures retrieved from occultation of a star by Neptune in 1981. For comparison, two extreme radiative-convective models are also shown. (French et al. 1983b)

file in the $10-900$ mb region on Uranus (Tyler et al. 1986; Hanel et al. 1986). The pertinent ground-based data are those at the microwave wavelengths (Gulkis and dePater 1984), and the ensuing infrared wavelengths (used in radiative transfer inversion): 8 to 14 µm, $16.5-22$ µm and $30-55$ µm (Courtin et al. 1978; Tokunaga et al. 1983; Orton et al. 1983; Mosely et al. 1985; Hildebrand et al. 1985; Orton et al. 1986; Aitken et al. 1986).

There is a striking difference between the energy balance of Uranus and Neptune. Uranus has a small internal heat source, equivalent to less than 15% of the absorbed solar energy. Neptune, on the other hand, may be emitting as much as three times the energy it absorbs from the Sun. The lack of substantial internal heat at Uranus may imply that at some level deep in the troposphere the atmosphere of Uranus could become isothermal, similar to the situation below the thermocline in the terrestrial ocean. Wallace (1980) has calculated that an internal energy flux of $\sim 100 \, \mathrm{erg \, cm^{-2} \, s^{-1}}$ would be needed to maintain an adiabatic profile below the radiative-convective boundary at ~ 1 bar. Since the strength of internal heat flux on Uranus appears comparable to $100 \, \mathrm{erg \, cm^{-2} \, s^{-1}}$, it would seem that an adiabatic thermal profile (modified by condensing clouds) is quite feasible in the troposphere. Bear in mind, however, that there has been some indication from the radio spectrum longward of 6-cm that Uranus may be becoming isothermal at a temperature of approximately 255 K (reached at a pressure of ≥ 40 bar), (Gulkis and dePater 1984). This interpretation is, however, not unique, and the likelihood of Uranus becoming isothermal deep in the troposphere is remote.

Both Uranus and Neptune attain a temperature of $75-78$ K at 1 bar, and 150 K at ~ 10 bar in the convective regime. The Voyager infrared and the radio occultation data yield a temperature of $74-76$ K at 900 mb on Uranus (Hanel et al. 1986; Tyler et al. 1986). The temperature minimum is reached at ~ 100 mb level, where the temperature is 52 K on the illuminated (south) pole, and 54 K on the dark (north) pole of Uranus. The tropopause region is extensive, from $80-110$ mb in the south polar region, and $60-200$ mb in the north polar region. These Voyager results are very similar to the ground-based data, which also indicate that the tropopause conditions at Neptune are nearly the same as on Uranus. The Voyager measurements of Uranus also show that there is virtually no pole-to-equator variation in the tropospheric temperature, either on the illuminated or the dark hemisphere. The only exception is a cold band (or collar) located between approximately 10° and 40° latitudes. The temperature minimum occurs at around 25° latitude on the illuminated hemisphere, and 40° latitude on the dark hemisphere, where the temperatures are lower by 1 to 2 K. Dynamical redistribution of the solar and internal heats must play a dominant role in maintaining the observed thermal behavior on Uranus, as the radiative time constant in the troposphere is long (7×10^9 s, or more than twice the sidereal orbital period). The temperature at 1 mb is around $88-90$ K (Voyager/Uranus Photopolarimeter data). The stratospheric temperature gradient in the 100 to 0.1 mb region is greater on Neptune than

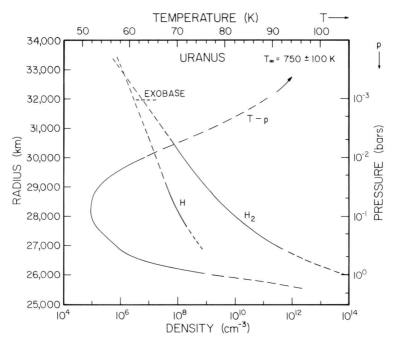

Fig. 2.20. Temperature and H_2 and H density profiles in the atmosphere of Uranus. *Broken lines* represent inter- and extrapolations of the Voyager measurements. Temperatures are plotted against atmospheric pressure (*T-p* curve). Densities are plotted against planetocentric distance (radius)

on Uranus. It could be due to the greater absorption by CH_4 which is supersaturated in the stratosphere of Neptune. Measurements indicate a stratospheric mole fraction of 2% for CH_4 on Neptune, the same as in the troposphere even though there is a cold trap in between (Orton et al. 1983). Strong vertical updrafts can cause migration of methane ice crystals from the cold trap to the lower stratosphere, where they would evaporate due to the warmer temperatures prevalent there (Hunten 1984). The stratospheric temperature structure is best explained by the radiative transfer models which include approximately 10 to 15% absorption of the solar radiation by aerosols or dust on Uranus, and 5% absorption on Neptune (Appleby 1986). Condensation of acetylene, polyacetylene, and ethane, which are produced in the photochemistry of CH_4 (Atreya and Ponthieu 1983; Atreya 1984a; see also Chap. 5 on Photochemistry) can supply the needed aerosols. The smaller absorption by aerosols required at Neptune is perhaps again due to the large supersaturated abundance of CH_4 in the stratosphere.

The stellar occultation observations in the visible yield temperatures in the $3-30$ µb region (Dunham et al. 1980; French et al. 1983a, b; Sicardy et al. 1982; Sicardy et al. 1985). The temperature around the 1-µb level is

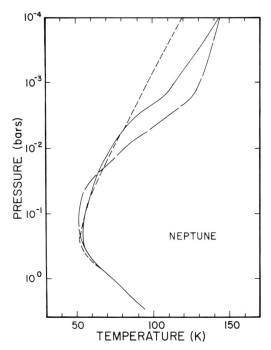

Fig. 2.21. Temperature structure of Neptune derived from the ground-based infrared data by Tokunaga et al. (1983), *solid lines*. Both *long-dashed line* radiative-convective model of Appleby (1986) and a *simple straight line* stratospheric model are consistent with the data. (Tokunaga et al. 1983)

120 – 160 K on both planets, although there are large (~ 30 K) variations and error-bars associated with these data.

The Voyager ultraviolet spectrometer monitored occultations of the Sun and stars γ Pegasi and ν Geminorum by Uranus. The γ Pegasi occultation took place at high latitudes (66.5° entrance, – 69° exit), whereas the occultations of the Sun and ν Geminorum occurred in the illuminated equatorial region (+ 2° to + 6°). A preliminary analysis of the best light curves, obtained from γ Pegasi and the solar occultations, yields a surprisingly high exospheric temperature of 750 ± 100 K at a radius of 27,720 km (Broadfoot et al. 1986). The light curves best suited for the measurement of exospheric temperature (and the H_2 density) are those where H_2 absorbs continuously ($\lambda < 845$ Å) and where it has its Lyman and Werner band absorptions (845 – 1150 Å). The temperature in the exosphere, much like that in the troposphere, is found to be independent of latitude or illumination conditions, implying strong dynamical redistribution of energy over the planet.

In addition to the H_2 density and the exospheric temperature, it is possible to derive the distribution of atomic hydrogen in the upper atmosphere. The light curves ideally suited for such a measurement are in the wavelength range 845 – 912 Å, as H absorbs continuously there while H_2 does not. The above-mentioned ultraviolet solar occultation experiment, when analyzed in this manner, yields a highly extensive corona of hydrogen around Uranus. In view of the 750 K exospheric temperature, it is not entirely surprising. A further clue to the existence of a corona was provided by the observation of a planet-wide glow of H and H_2 emissions, dubbed electroglow (further discussion on the electroglow phenomena can be found in Chap. 4 on Vertical Mixing). Electroglow on Uranus is initiated by soft electrons, which are also expected to produce a large population of fast H atoms (on dissociation of H_2) with energy in excess of the escape energy. The resulting Uranus corona of hydrogen is similar to the geocorona. It should therefore be possible to define an "exobase" or "critical level" (Appendix A.2.4), and derive the density distribution in the exosphere using the barometric law (Appendix A.2.4). The exobase is calculated to be at a planetocentric distance of 1.25 R_U (1 R_U = 25,550 km), where the atmospheric density is 6×10^6 cm^{-3}. A model of the atmosphere of Uranus based on the above-mentioned measurements is shown in Fig. 2.20. Stellar, solar, and radio occultations are also planned at Neptune when Voyager arrives there in August, 1989. The best currently available thermal structure of Neptune can be described by the curves shown in Fig. 2.21.

2.6 Outstanding Issues

What is the thermal structure in the middle atmospheres of the major planets? This serious gap in information on the upper stratosphere and the mesosphere can be alleviated by pre-entry probe measurements. Carefully designed stellar and solar occultations, as well as airglow measurements at long ultraviolet wavelengths (1500 – 2500 Å) can help also.

Are there temporal and latitudinal changes in the thermospheric heat balance? This requires ultraviolet and radio occultation observations from orbiters.

Why is the energy balance at Neptune so much greater than that at Uranus, particularly in view of the fact that the physical characteristics of the two planets are so similar? Do aerosols, hazes, and clouds significantly affect the tropospheric thermal models? Some of these issues are expected to be addressed by Voyager and Galileo spacecrafts, while others will still remain unresolved. To accomplish this, additional ground-based data in the microwave and radio range, and in-situ measurements from deep atmospheric probes will be needed.

A.2 Appendix

A.2.1 Great Red Spot

1. A number of cloud features, ovals, barges, ribbons, plumes, etc. have been seen in the images of Jupiter and Saturn taken by Voyager (see Figs. 2.22 and 2.23). Plumes have been imaged by Voyager in the mid-high latitudes of Uranus, and cloud features are believed to exist on Neptune as well. The most out-

Fig. 2.22. A whole disc image of Jupiter taken 1 month prior to the Voyager 1 encounter with the planet on 5 March 1979. The resolution is 525 km per line pair, which is very nearly the best available from Voyager. Many meteorologically prominent features are evident in this image. To the south of the equator are the Great Red Spot and the relatively young (~ 50 years old) white oval clouds (at $-40°$). To the north of the equator are a white plume, and three brown spots (midlatitude region). In addition, several alternating eastward-moving (V_E) and westward-moving (V_W) currents can be noticed parallel to the equator. These are located at latitudes $50°$ (V_w), $47°$ (V_E), $44°$ (V_w), $41°$ (V_E), $38°$ (V_w), $35°$(V_E), $30°$ (V_w), $23°$ (V_E), $18°$ (V_w), $8°$ (V_E), $-8°$ (V_E), $-16°$ (V_w), $-23°$ (V_E), $-28°$ (V_w), $-32°$ (V_E) and $-35°$ (V_w). Two of the Galilean satellites are also seen in this image: Io crossing to the east, and Europa, off the bright limb. (Courtesy of NASA/Voyager Project)

standing of these features is the GRS, which has been observed for at least 300 years, and has been around perhaps even longer. The GRS is centered around a zenographic latitude of approximately 23°S, i.e., in the South Tropical Zone. It spans over a 10° latitude range, with approximate dimensions of 13,000 km (width) and 26,000 km (length). The GRS is an anticyclonic feature in an anticyclonic shear zone. The period of rotation within the GRS is about 6 days (from Voyager imaging data of Smith et al. 1979a, b) which is consistent with the eastward wind velocity 75 m s^{-1} and north-south wind velocity of 40 to 50 m s^{-1}, as derived from the horizontal temperature gradients (Voyager infrared data of Hanel et al. 1979a, b).

Fig. 2.23. An image of the whole disc of Saturn and its rings taken 3 weeks prior to the arrival of Voyager 2 at Saturn on 26 August 1981. Although Saturn does not possess as many meteorological features (such as spots, clouds, ovals, barges, etc.) as Jupiter, it is nonetheless extremely active meteorologically. The current systems quite prominent in this image are the dark North Equatorial Belt (NEB) at 20°N, and the more active North Temperate Belt (NTeB) at 40°N. In the high resolution image (not shown) taken with the Voyager narrow angle camera, many alternating zonal jets are also seen, most prominent of which are the eastward-flowing jet at 60°N, westward at 55°N and eastward at 50°N. In this figure, three of Saturn's moons, Tethys, Dione and Rhea, are seen as three bright spots below the southern hemisphere. Also seen is the shadow of Tethys on Saturn. The A, B, and C rings are prominently displayed in the decreasing order of planetocentric distance. The gap in the A ring (called Encke or Keeler division) and that between the A and the B rings (called Cassini division) are equally prominent in this image.

A.2.2 Equatorial Belts and Zones

Both Jupiter and Saturn possess a number of distinct current systems flowing with westward and eastward velocities parallel to the equator (see Figs. 2.22 and 2.23). Bands similar to those on Jupiter and Saturn have also been seen in the Voyager images of Uranus. They are presumably present on Neptune also. These currents can be likened to the westward-flowing trade winds at low latitudes, and the eastward-flowing jet stream at mid to high latitudes in the Earth's atmosphere. There are about six currents of each kind at Jupiter, with typical eastward equatorial velocities of 50 to $100 \, \text{m s}^{-1}$ (Smith et al. 1979a, b). Saturn has fewer but higher velocity currents with eastward zonal velocities at the equator as high as $500 \, \text{m s}^{-1}$ relative to the planetary spin rate (Smith et al. 1981). The bright and dark bands parallel to the equator are referred to as zones and belts, respectively. Their latitudinal boundaries are virtually invariant temporally. The zones are believed to be regions of rising air (with associated cooler temperatures). Their bright appearance is perhaps due to the whitish ammonia clouds. The belts, on the other hand, represent regions of sinking air with associated warmer temperatures. Their visually dark appearance may result from the relative absence of cloud particles.

A.2.3 Line Broadenings

The resonance absorption or emission lines are broadened due to the finite lifetime of the excited states (natural broadening), velocity of the emitters relative to rest frame (Doppler broadening), and the pressure of the excited species or of the background gas (pressure or Lorentz broadening). The expressions for the full width at half maximum (FWHM) of a line due to these broadenings are discussed below.

Using Heisenberg's principle of uncertainty, the following expression for the natural broadening is obtained,

$$\alpha_N [\text{cm}^{-1}] = \frac{1}{2 \pi c \tau} = \frac{5.3 \times 10^{-12}}{\tau} ,$$

where c is the velocity of light in cm s^{-1}, and τ is the lifetime of the excited species in s. α_N is generally on the order of $10^{-3} \, \text{cm}^{-1}$, i.e., a factor of 100 to 1000 lower than the instrument width of high-resolution spectrometers.

In Doppler broadening, the absorbed or emitted lines appear broadened because the motion of the absorber/emitter relative to rest results in the shift of frequency by the following amount

$$\Delta \nu = \nu_0 \frac{\nu_1}{c} ,$$

where v_l is the line of sight velocity of the emitters. With Maxwell-Boltzmann distribution of velocities, the expression for Doppler broadening turns out to be

$$a_D \, [\text{cm}^{-1}] = 2 \sqrt{2 \, R \ln 2} \, \frac{v_0}{c} \sqrt{\frac{T}{M}} \, ,$$

where R is the gas constant, T is the temperature (in K), and M is the molecular weight (in AMU) of the emitter, and v_0 is the line center frequency (in cm^{-1}).

Thus,

$$\alpha_D \, [\text{cm}^{-1}] = 7.16 \times 10^{-7} \, v_0 \sqrt{\frac{T}{M}}$$

or

$$\alpha_D \, [\text{Å}] = 7.16 \times 10^{-7} \, \lambda_0 \sqrt{\frac{T}{M}} \, , \, \lambda_0 \, [\text{Å}] \, .$$

For hydrogen Lyman-α (1216 Å), e.g., the typical value of α_D at 1000 K is 0.0275 Å (1.86 cm^{-1}), which is comparable to the best available high-resolution instrument widths.

Finally, line broadening occurs also when the pressure of the absorbing gas is greater than ~ 0.01 mb, or when the background gas pressure exceeds a few millibars. Collision of absorbing atoms with gas of the same kind results in self- or Holtsmark broadening, while collisions with background gas result in Lorentz broadening. The Lorentz width is given by the following expression

$$\alpha_L \, [\text{cm}^{-1}] = N_L / c \pi \, ,$$

where N_L is the number of collisions per atom per s with the background gas. α_L increases linearly with the background pressure. For most atmospheric species, α_L and α_D are roughly comparable at 100 mb and 300 K.

A combination of natural, Doppler, and pressure broadening yields the Voigt profile of the line.

A.2.4 Critical Level

For light gases that are likely to escape from the gravitational field of a planet, critical level marks a significant interface between the thermosphere below and the exosphere above. This level is also referred to as the exobase. At the critical level a high velocity particle (atom or molecule) with Maxwell distribution has a 37% (i.e., e^{-1}) probability of either escaping without undergoing additional collisions with the background gas, or ending up in an elliptical orbit (Jeans 1954). The altitude of the critical level can be calculated by equating the mean free path for collisions between the light particles and the background atmosphere, λ_c, to the atmospheric scale height, H_c at the critical level (Chamberlain 1963), i.e.,

$$\lambda_c = (Q n_c)^{-1} = H_c,$$

where Q is the collision cross-section ($\sim 3 \times 10^{-15} \, \text{cm}^2$), and n_c is the total number density at the critical level. As is evident from the above expression, the mean free path increases with height.

In the exosphere the temperature remains constant with altitude. Therefore, the height (or radial) distribution of exospheric constituents can be calculated by employing the barometric equation (see Chap. 4 for additional details), i.e.,

$$n(R_2) = n(R_1) \exp \left[- \int_{R_1}^{R_2} dR/H(R) \right],$$

where $n(R_1)$ and $n(R_2)$ are, respectively, particle densities at radius R_1 and R_2, and $H(R)$ is the particle scale height at radius R, i.e.,

$$H = \frac{kT}{mg}.$$

Considering the variation of gravity with radius,

$$g(R) = g(R_1)(R_1/R)^2,$$

the above-mentioned expression for the radial distribution of the exospheric constituent reduces to

$$n(R_2) = n(R_1) \exp \left[- \left(\frac{R_1}{H(R_1)} - \frac{R_2}{H(R_2)} \right) \right].$$

R_1 is generally taken to be the radius of the exobase.

3 Cloud Structure

The cold temperatures prevalent in the tropospheres of the outer planets can cause condensation of numerous atmospheric gases, resulting in a multilayered cloud structure. On Jupiter, for example, the predicted clouds in the region of the Galileo/Probe descent are ammonia ice (NH_3-ice), ammonium hydrosulfide crystals (NH_4SH-s), and water-ice (H_2O-ice) as illustrated in Fig. 1.5. On Uranus and Neptune, clouds of methane ice (CH_4-ice) and aqueous ammonia ($NH_3 - H_2O$ solution) are also likely, in addition to the above-mentioned cloud layers. Ground-based as well as Pioneer and Voyager observations of Jupiter and Saturn are consistent with the topmost cloud as made up of NH_3-ice (see Figs. 2.22 and 2.23 for the observed cloud forms). The existence and nature of the other clouds are at this time simply postulated on theoretical grounds. Preliminary numerical simulation of the cloud structure on the major planets was done by Weidenschilling and Lewis (1973), whereas the results of the most recent calculations have been published by Atreya and Romani (1985).

3.1 Lapse Rates, Cloud Densities

The basic characteristics of the outer planet clouds can be derived from simple mathematical considerations assuming thermodynamic equilibrium. Whenever the partial pressure of a constituent exceeds its saturated vapor pressure, condensation occurs. This process releases latent heat of condensation, thereby changing the local lapse rate. It is reasonable to assume that in the region of the cloud formation, the atmospheric thermal structure is governed by a wet adiabat. An expression for the wet adiabatic lapse rate (dT/dz) at Jupiter can be derived using the law of energy conservation for adiabatic expansion of one mole of gas, i.e.,

$$\bar{C}_p dT - v dP + \sum_k L_k dX_k + L_{RX} dX_{H_2S} = 0 , \tag{3.1}$$

where $\bar{C}_p$ is the mean molar heat capacity at constant pressure, dT is the differential change in absolute temperature, v is the molar volume, dP is the differential change in total pressure, L_k is the molar enthalpy of condensation of the kth gas, dX_k is the differential change in the number of moles of the kth gas due to condensation, L_{RX} is the molar heat of reaction for:

$$NH_3(g) + H_2S(g) \rightarrow NH_4SH(s) \tag{3.2}$$

(this reaction represents the only cold trap for H_2S, which otherwise cannot condense anywhere on Jupiter because of its high vapor pressure), and dX_{H_2S} is the differential change in the number of moles of H_2S due to the reaction in Eq. (3.2). Since in the above reaction for every mole of hydrogen sulfide one mole of ammonia is removed, the choice of dX_{H_2S} in Eq. (3.1) is arbitrary, i.e., dX_{NH_3} and dX_{H_2S} are interchangeable. Using the standard expression for hydrostatic equilibrium, the fact that for an ideal gas mixture $X_k = P_k/P$ (where P_k is the partial pressure of the kth gas), the Clausius-Clapeyron equation,

$$dP_k/dT = P_k L_k / RT^2, \tag{3.3}$$

and the following expression for equilibrium constant, K_p, (Lewis 1969) of the reaction given by Eq. (3.2)

$$\log_{10}(K_p) = \log_{10}(P_{NH_3} \cdot P_{H_2S}) = 14.82 - 4705/T, \tag{3.4}$$

where P is expressed in atmospheres, and T in K, one can obtain from Eq. (3.1) the following expression for the wet adiabatic lapse rate (Atreya and Romani 1985; Weidenschilling and Lewis 1973).

$$\frac{dT}{dz} = \frac{-\bar{m}g}{\bar{C}_p} \frac{\left[1 + \dfrac{1}{RT}\left(\sum_k L_k X_k + \dfrac{2(X_{H2S} \cdot X_{NH3})L_{RX}}{(X_{H2S} + X_{NH3})}\right)\right]}{\left[1 + \dfrac{1}{\bar{C}_p T^2}\left(\sum_k \dfrac{L_k^2 X_k}{R} + \dfrac{(X_{H2S} \cdot X_{NH3})}{(X_{H2S} + X_{NH3})} \cdot L_{RX} \cdot 10{,}834\right)\right]}, \tag{3.5}$$

where $\bar{m}$ is the atmospheric mean molecular weight. On Uranus and Neptune (and to some extent, Saturn), one other possible condensate is aqueous ammonia solution of variable concentration. This requires modification of the $L_k dX_k$ term in the above expression. If the solution that condenses out is of concentration C (in mole fraction of ammonia), then for every mole of solution, $1-C$ moles of H_2O condense out, i.e.,

$$dX_{solution} = \frac{dX_{H_2O}}{1-C}. \tag{3.6}$$

Thus, the $L_k dX_k$ term in Eq. (3.5) due to the formation of a solution cloud can be expressed as

$$\frac{L_s dX_{H_2O}}{1-C}, \tag{3.7}$$

where L_s is the average heat of condensation of the solution. Even though the heat of condensation of the aqueous ammonia solution is a function of both the temperature and the concentration of the solution, Lewis (1969) showed that it is permissible to adopt an average heat of condensation for the solution cloud. In the outer planets, for various physical reasons, the solution cloud becomes more concentrated in ammonia as the temperature decreases. These two factors act to keep the heat of condensation approximately constant. At any rate, the heat of condensation varies by less than a factor of 2. Using the hydrostatic law, the following expression can be obtained for the mean cloud density ($\bar{D}$) of the kth condensate between two closely spaced atmospheric levels I and J (Weidenschilling and Lewis 1973).

$$\bar{D} = \frac{m_k(X_k^I - X_k^J)\bar{P}}{\bar{m}gdz} , \tag{3.8}$$

where m_k is the mass fraction of the kth condensate, P is the mean atmospheric pressure between I and J levels, and dz is the height increment from I to J levels. If more than one species is condensing to form the cloud (NH$_4$SH or aqueous ammonia cloud), the right-hand side of Eq. (3.8) is summed over the condensing species.

3.2 Thermodynamic Data

The presently available data on the heat capacities and vapor pressures of the condensibles most likely to be encountered on the major planets are summarized below.

3.2.1 Heat Capacities and Latent Heats

The values of the molar heat capacity at constant pressure, C_p, and the latent heats of the condensibles in the outer planets are given in Table 3.1. The average dry adiabatic lapse rates, Γ_d, relevant for the midlatitudes of Jupiter, Saturn, Uranus, and Neptune are, respectively, 1.91, 0.8, 0.7, and 0.85 K km^{-1}, assuming solar composition atmospheres. If the He/H$_2$ ratio differs substantially from solar in the region of clouds, then Γ_d would be different also, e.g., at Saturn it would decrease to 0.75 K km^{-1} with He/H$_2$ = 3.5% by volume.

3.2.2 Vapor Pressures

The behavior of the saturation vapor pressures of CH$_4$, C$_2$H$_4$, C$_2$H$_6$, C$_2$H$_2$, H$_2$S, NH$_3$, and H$_2$O with temperature is shown in Fig. 3.1. Empirical relationships for the vapor pressures are given below.

H$_2$O-*vapor over* H$_2$O-*ice (Washburn 1924)*

$$\log_{10} V_p = -\frac{2445.5646}{T} + 8.2312 \log_{10} T - 0.01677006\, T$$
$$+ 1.20514 \times 10^{-5}\, T^2 - 6.757169 \tag{3.9}$$

(for 173 K $\leq T \leq$ 273 K)

Table 3.1. Mean molar heat capacities and latent heats

Substance	C_p (at room temperature) $[10^8 \text{ erg mol}^{-1} \text{ K}^{-1}]$	L^a $[10^{11} \text{ erg mol}^{-1}]$
H_2	2.877	
He	2.081	
H_2O	3.341	5.105 (subl)
CH_4	3.572	1.024 (subl)
NH_3	3.707	2.280 (subl)
H_2S	3.338	2.200 (subl)
Ar	2.074	0.763 (subl)
Aqueous-NH_3 solution		4.000 (vap)
NH_4SH-solid		9.312 (form)

[a] subl: sublimation; vap: vaporization; form: heat of formation.

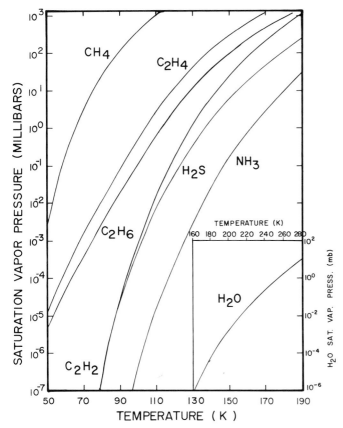

Fig. 3.1. Saturation vapor pressures of the condensible gases on the outer planets. *Inset* shows the H_2O saturation vapor pressure behavior with temperature which is given on the upper abscissa of the inset

V_p is the vapor pressure in mmHg, and T is the temperature in K.

NH$_3$-*vapor over* NH$_3$-*ice (Karwat 1924)*

$$\log_{10} V_p = -\frac{1790}{T} - 1.81630 \log_{10} T + 14.97593 \tag{3.10}$$

(for 100 K $\leq T \leq$ 195 K),

V_p is in mmHg.

NH$_3$-*vapor over* NH$_3$-*liquid (International Critical Tables 1928)*

$$\log_{10} V_p = 27.376004 - \frac{1914.9569}{T} - 8.4598324 \log_{10} T + 2.39309 \times 10^{-3} T$$
$$+ 2.955214 \times 10^{-6} T^2 \tag{3.11}$$

(for 195 K $\leq T \leq$ 343 K).

V_p is in atmospheres.

H$_2$S-*vapor over* H$_2$S-*ice (Giauque and Blue 1936)*

$$\log_{10} V_p = -\frac{1329}{T} + 9.28588 - 0.0051263 \, T \tag{3.12}$$

(for 164.9 K $\leq T \leq$ 187.6 K).

V_p is in cmHg.

H$_2$S-*vapor over* H$_2$S-*liquid (Giauque and Blue 1936)*

$$\log_{10} V_p = -\frac{1145}{T} + 7.94746 - 0.00322 \, T \tag{3.13}$$

(for 187.6 K $\leq T \leq$ 213.2 K).

V_p is in cmHg.

CH$_4$-*vapor over* CH$_4$-*ice (Ziegler 1959)*

$$\log_{10} V_p = 7.69540 - \frac{532.20}{T + 1.842} \tag{3.14}$$

(for 50 K $\leq T \leq$ 90 K).

V_p is in mmHg.

C$_2$H$_2$-*vapor over* C$_2$H$_2$-*ice (Ziegler 1959)*

$$\log_{10} V_p = 9.130 - \frac{1149.0}{T - 3.840} \tag{3.15}$$

(for 115 K $\leq T \leq$ 145 K),   and

$$\log_{10} V_p = -8.6252 \log_{10} T - \frac{1.6459 \times 10^3}{T} + 3.0895 \times 10^1 \qquad (3.16)$$

(for $98.6 \text{ K} \le T \le 115 \text{ K}$).

V_p is in mmHg.

C_2H_4-*vapor over* C_2H_4 *(Ziegler 1959)*

$$\log_{10} V_p = 6.74756 - \frac{585.00}{T - 18.16} \qquad (3.17)$$

(for $120 \text{ K} \le T \le 155 \text{ K}$),   and

$$\log_{10} V_p = -17.14 \log_{10} T - \frac{1703}{T} + 50.79 \qquad (3.18)$$

(for $104 \text{ K} \le T \le 120 \text{ K}$),   and

$$\ln V_p = -4.3732 \ln T - \frac{2372.0}{T} + 43.024 \qquad (3.19)$$

(for $T < 104 \text{ K}$).

The triple point of C_2H_4 is 104 K. Use the first two expressions for vapor over liquid. The last one may be used for vapor over ice; however, the following are more appropriate to use.

$$\log_{10} V_p = 8.724 - \frac{901.6}{T - 2.555} \qquad (3.20)$$

(for $89 \text{ K} \le T \le 104 \text{ K}$),   and

$$\log_{10}(V_p \times 10^3) = 1.5477 - 1038.1\,(1/T - 0.0110)$$
$$+ 16537\,(1/T - 0.0110)^2 \qquad (3.21)$$

(for $T < 89 \text{ K}$)

V_p is in mmHg in all above vapor pressure relations for C_2H_4.

C_2H_6-*vapor over* C_2H_6 *(Ziegler 1959)*

$$\log_{10} V_p = 6.80266 - \frac{656.40}{T - 17.16} \qquad (3.22)$$

(for $135 \text{ K} \le T \le 155 \text{ K}$),   and

$$\log_{10} V_p = 9.172 - \frac{1196}{T + 15.69} \qquad (3.23)$$

(for $90 \text{ K} \le T \le 135 \text{ K}$),   and

$$\ln V_p = -4.2133 \ln T - \frac{2459}{T} + 41.32 \qquad (3.24)$$

(for $T < 90$ K).

The triple point of C_2H_6 is 90 K. Use the first two expressions for vapor over liquid; the last may be used for vapor over ice. However, for 83 to 89 K, the following expression should be used.

$$\log_{10} V_p = 8.489 - \frac{957.7}{T} \qquad (3.25)$$

(for $83 \text{ K} \le T \le 89 \text{ K}$).

V_p is in mmHg in all above vapor pressure relations for C_2H_6.

Ar-*vapor over* Ar-*ice (Stull 1947)*

$$\log_{10} V_p = 6.204 - \frac{237.0}{T - 16.80} \qquad (3.26)$$

(for $55 \text{ K} \le T \le 82.6 \text{ K}$).

V_p in mmHg.

HCN-*vapor over* HCN-*ice (Stull 1947)*

$$\log_{10} V_p = 8.6165 - \frac{1516.5}{T - 26.20} \qquad (3.27)$$

(for $202 \text{ K} \le T \le 260 \text{ K}$).

V_p is in mmHg.

HCN-*vapor over* HCN-*liquid (Stull 1947)*

$$\log_{10} V_p = 9.028 - \frac{2225.4}{T + 62.91} \qquad (3.28)$$

(for $260 \text{ K} \le T \le 305 \text{ K}$).

V_p is in mmHg.

Note that melting point of HCN is 260 K.

N_2-*vapor over* N_2-*ice (Giauque and Clayton 1933)*

$$\log_{10} V_p = (-381.6/T) + 7.41105 - 0.0062372 \, T \qquad (3.29)$$

(for $54.78 \text{ K} < T < 63.14 \text{ K}$)

N_2-*vapor over* N_2-*liquid (Giauque and Clayton 1933)*

$$\log_{10} V_p = (-339.8/T) + 6.71057 - 0.0056286 \, T \qquad (3.30)$$

(for $63.14 \text{ K} < T < 78.01 \text{ K}$)

V_p is in cmHg in the above expressions for N_2.

The saturation vapor pressures of H_2O and NH_3 over aqueous ammonia solution are available in the range $273 - 363$ K (Wilson 1925, Linke 1965; International Critical Tables 1928). However, on Uranus, Neptune, and perhaps Saturn, the solution cloud may form at even lower temperatures. The lack of available data at lower temperatures poses a serious problem. Neither the use of the Clausius-Clapeyron equation (Atreya and Romani 1985), nor polynomial fitting (Weidenschilling and Lewis 1973) to the available data for extrapolation to lower temperatures is strictly valid in this situation. The former, however, seems to be physically more realistic.

3.3 Cloud Models

The cloud density calculations are initiated at some level deep in the troposphere where condensation has not occurred. The temperature varies with altitude according to the dry adiabatic lapse rate. The altitude is incremented in small steps. When the partial pressure of a constituent exceeds its equilibrium saturation vapor pressure, or in the case of the NH_4SH cloud the equilibrium constant is exceeded, condensation is presumed to occur. The condensate density and the new lapse rate are then calculated according to the formulations presented earlier. The resulting typical cloud structures at Jupiter, Saturn, and Uranus are shown in Figs. 3.2, 3.3, and 3.4.

Since the temperature at the 1-bar level is the same at both Uranus and Neptune, the thermochemical equilibrium cloud structure on the temperature scale is expected to be virtually identical for these two planets, assuming similar composition for both.

For Jupiter, an extreme situation with H_2O mole fraction of 10^{-6} is also shown since it represents the lowest O/H ratio measured so far at Jupiter (see Chap. 1 on Composition for additional details). The resulting H_2O-ice cloud is virtually nonexistent, as is evident from Fig. 3.2. Many important questions including possible modification of thermochemical equilibrium cloud structure by dynamics and the extent and density of water cloud on Jupiter will be addressed by the nephelometer and the mass spectrometer on the Galileo Entry Probe. The Probe is designed to operate nominally down to a pressure level of 10 bars; however, information is possible down to approximately the 20-bar level.

The relatively large densities of Uranus and Neptune imply greater than solar mixing ratios of the volatiles (see Chap. 1 on Composition for additional details). The only molecule whose vapor-phase tropospheric abundance can be presently determined with some degree of confidence is CH_4. From various analyses of the thermal radiance and geometric albedo data, it is concluded that the CH_4 mole fraction on Uranus and Neptune is at least 2 to 3%, or 20 to 40 times solar (Orton et al. 1986; Pollack et al. 1985; Bergstralh and Baines 1984). The effect of enhanced CH_4, NH_3, H_2S, and H_2O is largely reflected in a greatly extended aqueous-NH_3 solution cloud, as seen in Fig. 3.4. In a solar

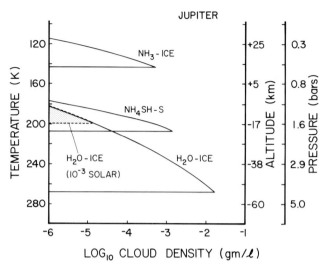

Fig. 3.2. A model of the low-latitude cloud structure at Jupiter. Solar elemental ratios are assumed for He, N, S, and C. The H_2O-ice cloud is shown for both solar and 10^{-3} solar (*shaded curve*) O/H ratios. Upper atmospheric thermal structure is from the Voyager radio science data. Pressures and altitudes on the right ordinates correspond to temperatures on the left ordinate. Altitudes are relative to the 1-bar pressure level. Calculations were begun deep in the atmosphere, where $T = 420$ K, $P = 20$ bar. The Voyager radio and infrared data indicate that the NH_3-ice cloud tops are at 0.5 to 0.6-bar level where the temperature is ~ 150 K

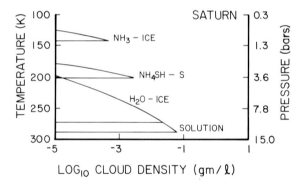

Fig. 3.3. Same as Fig. 3.2, except for Saturn. Here, $He/H_2 = 3.5\%$, $CH_4/H_2 = 0.45\%$, and all other volatiles have solar elemental ratios. Observations place NH_3 cloud tops at 0.6 to 0.7-bar level where the temperature is around 119 K

composition atmosphere, not all of the NH_3 is removed in the solution cloud. Of the remaining, some is removed by H_2S, resulting in the formation of an NH_4SH cloud, while that still left over forms a thin NH_3-ice cloud (see Fig. 3.4). In the more likely situation where the volatiles are enriched compared to their solar proportions, most of the NH_3 is lost in the extensive solution cloud, and the small amount of NH_3 which may still remain is entirely removed by H_2S to form the NH_4SH cloud. In fact, for all volatile enhancements of ten times or more over solar, there would be residual H_2S after the

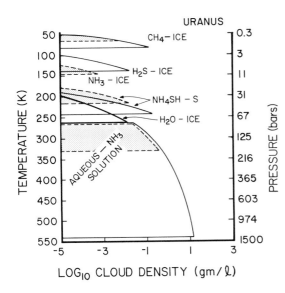

Fig. 3.4. A model of the cloud structure at Uranus with solar (*shaded curves*) and $10 \times$ solar ratios of the elements C, N, S, and O. The H_2O cloud forms at the same temperature in both cases. In the model with enhanced volatiles, a relatively tenuous and thin H_2S-ice cloud forms at around 140 K, and a substantial CH_4-ice cloud begins to form at approximately 1.5, 1.3, 1.1, and 0.55-bar level, respectively, for $30 \times$, $20 \times$, $10 \times$, and $1 \times$ solar C/H ratios. The currently favored value ($20-30 \times$ solar) places the CH_4 cloud base between 1.5- and 1.3-bar pressure level

Table 3.2. Ammonia mixing ratios by volume on Uranus[a]

T [K]	Solar[c]	Cold[b] 10 × Solar[d]	10 × H_2O[e]	Warm[b] Solar	10 × Solar
125	1.2×10^{-7}	0	0	3.3×10^{-7}	1.4×10^{-7}
150	4.0×10^{-6}	0	0	9.2×10^{-6}	1.4×10^{-7}
175	4.0×10^{-6}	3.6×10^{-11}	1.0×10^{-10}	9.2×10^{-6}	2.0×10^{-7}
200	4.2×10^{-6}	3.2×10^{-8}	9.4×10^{-8}	1.0×10^{-5}	3.0×10^{-6}
225	1.5×10^{-5}	6.2×10^{-6}	8.9×10^{-7}	2.2×10^{-5}	4.4×10^{-5}

[a] After Atreya and Romani (1985).
[b] Cold model has $T = 450$ K at $P = 670$ bar; Warm model has $T = 450$ K at $P = 250$ bar.
[c] Solar case has O/H, C/H, N/H, and S/H in the solar ratios of the elements.
[d] $10 \times$ solar case has O/H, C/H, N/H, and S/H in $10 \times$ solar ratios of the elements. NH_3 mixing ratios are even lower if N/H is assumed $30 \times$ solar.
[e] $10 \times H_2O$ case has only O/H = $10 \times$ solar; C/H, N/H, and S/H are in the solar ratios. This case is for illustration purposes only, and should not be constructed to imply author's prejudice.

formation of an NH_4SH cloud (Romani and Atreya 1985). This H_2S would result in the formation of an H_2S-ice cloud as shown in Fig. 3.4. In any event, the extent of NH_3 loss in the aqueous-solution cloud at Uranus and Neptune reflecting the solar or enhanced volatile ratios is so great that it would result in an apparent depletion of this species in the 125 to 250 K range, as noted in Table 3.2. The millimeter and centimeter observations of Uranus have been interpreted to imply $NH_3/H_2 < 10^{-6}$ in this temperature range (see review by Gulkis and dePater 1984). The extensive solution cloud in the above model would give the appearance of an ocean. Since the density of this "ocean" is so puny (<1 gm l^{-1}) compared to the terrestrial ocean, it is really more like a moist-vapor cloud. There is some likelihood of a hot water ocean surrounding the core, as discussed in Chapter 1. If such an ocean were present, it would be distinctly different and physically separated from the aqueous-ammonia solution.

The cloud structures of the major planets can be modified by the heat released in the conversion of para-hydrogen (antiparallel proton spin, $J = 0, 2, 4, \ldots$) to orthohydrogen (parallel proton spin, $J = 1, 3, 5, \ldots$). The "normal" ortho to para hydrogen ratio on the major planets is $3:1$, the high temperature limit (Herzberg 1966; Smith 1978). When H_2 is cooled to lower temperatures, the above ratio would be normally maintained since transitions involving $\Delta J = 1$ are quantum mechanically forbidden. However, above the 1-bar level on Jupiter, partial equilibration between the ortho and para states can occur in the presence of magnetic fields (Gierasch 1983), reactions with H-atoms and paramagnetic gases (Massie and Hunten 1982), and catalysis on aerosol particles (Massie and Hunten 1982). The extent to which equilibration might occur is not known. However, for equilibration to occur, the equilibration rate must be faster than the radiative time constant. The estimated change in the lapse rate is $\sim 10\%$, and the corresponding change in the temperature, ΔT, is ~ 0.5 K at approximately the 1-bar level on Jupiter (Massie and Hunten 1982). Although such small changes do not substantially alter the Jovian cloud structure, they are potentially quite significant in producing buoyancy contrasts and other dynamical effects in the troposphere (Gierasch 1983). The thermodynamical effects of the ortho-para hydrogen conversion could be more dramatic on the other outer planets, e.g., calculations of Massie and Hunten indicate ΔT of around 3 K at the 1-bar level on Saturn, and 1 K at the 1-bar level and ~ 3 K at the 8-bar level on Uranus. The resulting change in the lapse rate at the 10-bar level at Uranus would be approximately 0.2 K km^{-1}.

In the preceding thermodynamic equilibrium cloud models it is assumed that condensation occurs as soon as the partial pressure of the substance exceeds its saturation vapor pressure. In fact, a high degree of supersaturation may be allowed before phase transformation could take place. For condensation to occur, certain condensation nuclei (called embryos) must be present. When the nuclei are of the same substance as the condensing substance, homogeneous nucleation occurs. In the context of terrestrial water clouds, small water droplets are sometimes already present in a large volume of water vapor. They serve as embryos for homogeneous nucleation. Some of these embryos attain a critical size at the critical vapor pressure, disrupting the equilibrium between them and the phase of surrounding vapor. This can trigger a spontaneous phase transformation in the surrounding vapor. In an exceptionally clear airmass, there is a paucity of water droplets (or ice crystals), and although the homogeneous nucleation proceeds, it does so sluggishly. This can result in a vapor which is several hundred percent supersaturated. Heterogeneous nucleation, on the other hand, occurs at a much lower supersaturation, hence it is a much more effective nucleation process in planetary atmospheres. Chemical nucleation – e.g., formation of NH_4SH-solid on gas-phase reaction between NH_3 and H_2S – is equally effective. In the heterogeneous nucleation process the embryos form on foreign substances, such as aerosols, dust, etc. which are referred to as cloud condensation nuclei (CCN). For ice nucleation two stages are likely: first, near the freezing point, nucleation of supercooled water droplets, and then spontaneous freezing of these droplets. Direct ice-nucleation can occur only at much colder temperatures.

Once the condensation embryo grows to a critical size, it could precipitate in the form of rain or snow. Particle movement, however, is governed by two major – sometimes competing – forces. They are sedimentation, or gravitational fall, and wind transport. If the resultant fall-out time for the particle is shorter than its evaporation time, precipitation would take place. The terminal velocity, v, of the particle is given by the Stokes formula:

$$v = 2 g \rho_c a_c^2/9 \eta, \qquad (3.31)$$

where g is acceleration due to gravity, ρ_c is particle density, a_c is particle radius, and η is the atmospheric viscosity. The above expression is strictly valid for only solid spheres. The NH_3, NH_4SH, and H_2O-ice clouds on Jupiter, for example, are expected to have cylindrical, rhomboidal, and other geometrical shapes. The classical Stokes expression should therefore be modified according to the applicable gas and inertial drag forces (which are expressed in terms of Reynolds number). The presently available laboratory data indicate that the clouds on Jupiter would produce snow typically on a time scale of a few hours (Rossow 1978). Much further work in the area of laboratory measurements, theoretical modeling, and actual in-situ observations remains to be done before one truly understands the nature of clouds on the major planets.

3.4 Outstanding Issues

The most central outstanding issue is the role of dynamics in determining the eventual cloud structure of the outer planets. Information is presently lacking on the distribution, density, and the nature of condensation nuclei or embryos, rates of nucleation and sedimentation (fall-out in the form of rain or snow) of cloud particles, dynamical transport (up-draft or down-draft), and effect of freezing in liquid clouds. Little, if any, information is available on the horizontal distribution of clouds and its correlation with horizontal wind structure. In addition, even for constructing the first-order equilibrium models, thermodynamic data (vapor pressures, heat capacities, enthalpies etc.) are either nonexistent, outdated, or not available at the appropriate temperatures. This is particularly true for binary and tertiary molecular mixtures. Improved laboratory measurements as well as in-situ observations will be needed to understand fully the structure of the clouds on the major planets.

4 Vertical Mixing

4.1 Diffusion Equation

The distribution of constituents in a planetary atmosphere can be obtained by solving the continuity equation of the following form,

$$\frac{d\phi_i}{dz} = P_i - L_i \, , \tag{4.1}$$

where P_i and L_i are respectively production and chemical loss rates per unit volume, and ϕ_i is the vertical flux of the ith constituent. ϕ_i obeys the diffusion equation,

$$\phi_i = n_i(w_D + w_K) \, , \tag{4.2}$$

where n_i is the number density of the constituent, and w_D and w_K are respectively its molecular and eddy diffusion velocities. The latter can be expressed in terms of the density and thermal gradients (Colegrove et al. 1966), so that

$$\phi_i = n_i \left[-D_i \left(\frac{1}{n_i} \frac{dn_i}{dz} + \frac{1}{H_i} + \frac{1}{T} \frac{dT}{dz} \right) - K \left(\frac{1}{n_i} \frac{dn_i}{dz} + \frac{1}{H_a} + \frac{1}{T} \frac{dT}{dz} \right) \right] . \tag{4.3}$$

In the above expression, D_i and K are respectively molecular and eddy diffusion coefficients, and H_i and H_a are the pressure scale heights of the constituent and the background atmosphere, respectively, i.e.,

$$H_i = RT/M_i g \quad \text{and} \quad H_a = RT/M_a g \, , \tag{4.4}$$

where M_i and M_a are respectively the molecular weights of the constituent and of the atmosphere.

The eddy diffusion coefficient is a parameter generally used to define mixing in atmospheric applications. To some extent, it is a quantitative representation of turbulent mixing in a planetary atmosphere. Turbulence, however, results from small-scale vertical motions, whereas the eddy diffusion coefficient is used as a convenient parameter in which are lumped together all vertical motions, small as well as large. Eddy mixing tends to homogenize the atmosphere, so that in the absence of chemistry, all constituents are distributed according to the mean atmospheric scale height. Molecular diffusion, on the other hand, tends to separate the species according to their own individual molecular weights. The atmospheric level at which the molecular diffusion coefficient is mathematically equal to the eddy diffusion coefficient is defined as the "homopause" for the ith constituent (in the literature, homopause is also referred to as turbopause). Above this altitude, molecular diffusion dominates.

Based on the above discussion, two limiting solutions exist for the constituent's density — one above and one below the homopause. Above the homopause, $D_i \gg K$ (so that K can be neglected), and if the constituent has zero vertical flux ($\phi_i = 0$, e.g., due to no escape), then Eq. (4.3) would yield a "diffusive equilibrium solution" (i.e., $w_D = 0$). The constituent number density is then given by

$$n_i(z) = n_i(z_0)(T_0/T)\exp\left(-\int_{z_0}^{z} dz/H_i\right) \tag{4.5}$$

or, for isothermal conditions

$$n_i \propto \exp(-z/H_i) \ , \tag{4.6}$$

i.e., the constituents are distributed according to their own individual molecular weights above the homopause. The time constant for reaching diffusive equilibrium, τ_D, is given by the following expression (Chapman and Cowling 1970; Colegrove et al. 1966).

$$\tau_D = H_a^2/D_i \ . \tag{4.7}$$

In the other extreme, the eddy diffusion dominates well below the homopause, so that D_i can be set equal to zero in Eq. (4.3). Again, if the vertical flux is neglected ($\phi_i = 0$), the following expression for the constituent density in a well mixed atmosphere is obtained.

$$n_i(z) = n_i(z_0)(T_0/T)\exp\left(-\int_{z_0}^{z} dz/H_a\right) \ , \tag{4.8}$$

which reduces to the following form for an isothermal atmosphere,

$$n_i \propto \exp(-z/H_a) \ , \tag{4.9}$$

i.e., below the homopause, all species have the same scale height as the background atmosphere. In other words, the constituents are "mixed" in the homosphere, their mixing ratios are constant provided that chemistry or transport is not important [the mixing time constant is expressed, analogous to Eq. (4.7), as $\tau_K = H_a^2/K$]. For a family of species (such as CH_4 and its photochemical products, C_2H_2, C_2H_4, C_2H_6, etc.), the sum total of all their mixing ratios remains constant in the homosphere. When transport is important (e.g., when H produced in the H_2 photochemistry flows down to the lower atmosphere), the downward flux becomes a finite nonzero quantity. It is generally equal to the net production rate at the top. The constituent distribution then is calculated by solving Eq. (4.3) numerically.

In reality, transition from the eddy-diffusion-controlled to the molecular-diffusion-controlled regime is rather gradual and smooth. Although molecular diffusion coefficient (and the thermal diffusion coefficient, which is negligible here) can be deduced from considerations of gas kinetic theory, theoretical formulation of the eddy diffusion coefficient is exceedingly complex and the results are susceptible to gross uncertainties. A crude estimate of K can be made based on the mixing length theory (see, e.g., Lindzen 1971). K can be expressed in terms of the velocity, v, of two air parcels which interchange and mix thoroughly with the background over a distance L, i.e.,

$$K \sim vL \ . \tag{4.10}$$

Although it is difficult to quantify the vertical velocity, a good estimate for the range of v according to Hunten (1975) is 1 cm s^{-1} (fairly common) to 10^4 cm s^{-1} (rare). L, on the other hand, can be safely assumed comparable to the scale height, except perhaps in the troposphere and the lower stratosphere, where it is the height above the boundary. With the above values of v and L, one should expect K to be on the order of $2 \times 10^6 \text{ cm}^2 \text{s}^{-1}$ at Jupiter and Saturn, although higher values, up to $10^{10} \text{ cm}^2 \text{s}^{-1}$, are not excluded by the mixing length theory. In photochemical and other aeronomical applications, knowledge of K is, however, critical. Therefore key experimental techniques for determining K on the Jovian planets are discussed somewhat later in this chapter. They are based on considerations of Lyman-alpha and He 584 Å emissions, and the CH_4 distribution.

4.2 Limiting Flux

For light gases (such as H and H_2 on Titan, H and He on Earth) the thermal rate of escape from the top of the atmosphere is large. It is, however, limited by the diffusive flux from below. This important concept was developed by Hunten (1973a and 1973b) and Hunten and Donahue (1976) by transforming the flux Eq. (4.3) in terms of the volume mixing ratio of constituent i,

$$f_i = n_i/n_a \tag{4.11}$$

(n_a is the atmospheric number density).

Logarithmic differentiation of Eq. (4.11) yields

$$\frac{1}{f_i} \frac{df_i}{dz} = \frac{1}{n_i} \frac{dn_i}{dz} - \frac{1}{n_a} \frac{dn_a}{dz} \ . \tag{4.12}$$

Substitution of Eq. (4.12) in Eq. (4.3) eventually yields

$$\phi_i = \phi_1 - (D_i + K) n_a \frac{df_i}{dz} \ , \tag{4.13}$$

where the limiting flux, ϕ_1, is

$$\phi_1 = - n_i D_i \left(\frac{1}{n_a} \frac{dn_a}{dz} + \frac{1}{H_i} + \frac{1}{T} \frac{dT}{dz} \right) , \tag{4.14}$$

which for small or no thermal gradients can be approximated to

$$\phi_1 = \frac{n_i D_i}{H_a} (1 - M_i/M_a) , \tag{4.15}$$

and, since $M_i \ll M_a$ for a light gas flowing through the background atmosphere,

$$\phi_1 \approx \frac{n_i D_i}{H_a} . \tag{4.16}$$

With D_i expressed as

$$D_i = b_i/n_a , \tag{4.17}$$

(where b_i is called the binary collision parameter, discussed later) one obtains

$$\phi_1 \approx \frac{b_i f_i}{H_a} . \tag{4.18}$$

Thus, the limiting flux depends only upon the mixing ratio of the constituent (generally taken at the homopause) and the atmospheric scale height.

For an upward flow to occur, $df_i/dz > 0$ in Eq. (4.13), i.e., the mixing ratio, f_i, must increase with altitude. This also implies [from Eq. (4.13)] that $\phi_i < \phi_1$. If $df_i/dz = 0$, then $\phi_i = \phi_1$; constant f_i with altitude, however, does not result in a flow. Finally, ϕ_i can exceed ϕ_1 only if $df_i/dz < 0$, which implies a decreasing mixing ratio with altitude − a condition which clearly prevents upward flow from taking place. Thus, ϕ_1 represents truly the limiting flux of the constituent flowing through the background atmosphere. For energetic light atoms or molecules, the escape flux is nearly equal to the value of the limiting flux. In the outer planetary systems, the concept of limiting flux is most useful for determining the maximum allowable escape flux of H and H_2 from the exobase of Titan, and perhaps even Triton.

4.3 Methods for Determining Eddy Diffusion Coefficient

The eddy diffusion coefficient can be measured, in general, by analyzing the height distribution of a suitable tracer gas, or by inverting certain airglow data (whose behavior depends on vertical mixing) through the radiative transfer equation. The most powerful techniques for determining eddy diffusion coefficient on the outer planets are based on the measurement of H Lyman-alpha intensity, He 584 Å intensity, and the distribution of heavier gases, such as

CH_4, in the upper atmosphere. In the following subsections these methods are discussed individually in the context of Jupiter and Saturn. A brief discussion for Uranus and Neptune follows immediately afterwards.

4.3.1 Lyman-Alpha

Lyman-alpha emission from the Jovian planets is generally the result of resonance scattering of the solar Lyman-α photons by the atmospheric hydrogen atoms. Since CH_4 is a strong absorber of the Lyman-α photons, only those hydrogen atoms which lie above the methane homopause contribute to the observed Lyman-α emission, as the concentration of CH_4 drops off rapidly above the homopause. This means that for large values of eddy diffusion coefficient (i.e., high homopause level), fewer H atoms would be available for resonance scattering, resulting in a lower planetary Lyman-α intensity. The converse would be true for small values of eddy diffusion coefficient. Thus, it should be possible to relate the observed Lyman-α intensity to the eddy diffusion coefficient at the homopause, K_h, (Hunten 1969; Wallace and Hunten 1973). Since the Lyman-α albedo varies roughly as the square root of the H column abundance, N_H, it is implied that N_H above the homopause would have an inverse functional dependence on K_h. To determine K_h quantitatively, Eq. (4.3) is solved. The downward flux of H is equivalent to its production rate in the ionosphere, along with any contribution from the CH_4 photolysis. The column production of H is balanced by its 3-body recombination loss,

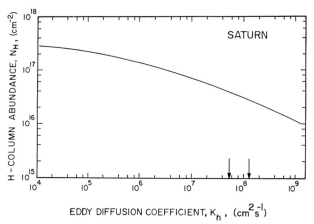

Fig. 4.1. Variation of the atomic hydrogen column abundance above the homopause of Saturn, N_H, with the homopause value of the eddy diffusion coefficient, K_h. *Arrows* represent K_h's corresponding to the N_H's deduced from the Voyager Lyman-α data interpreted with (*right arrow* larger K_h), and without (*left arrow* smaller K_h) the contribution of the interplanetary/interstellar Lyman-α. (After Waite 1981)

Table 4.1. Saturn Lyman-α

Observation date	Observation technique	Observed Ly-α intensity (kR)	Central disc Ly-α intensity (kR)	Reference
1975 Mar. 15	Rocket	0.70 ± 0.35	2.0 ± 1.0	1
1976 Apr. 15	Copernicus	0.45 ± 0.25	1.1 ± 0.6	2
1977 Apr. 28–30	Copernicus	0.80 ± 0.30	1.9 ± 0.7	2
1980 Jan. 19	IUE	0.8	2.1	3
1980 May 5	IUE	1.8 (auroral)	5.0 (auroral)	3
1980 Nov. 12	Voyager 1	3.3	3.3	4
1981 Aug.	Voyager 2	3.0	3.0	5

[1] Weiser et al. (1977); [2] Barker et al. (1980); [3] Clarke et al. (1981a); [4] Broadfoot et al. (1981a); [5] Sandel et al. (1982b).

following the transport of H atoms to the region below the homopause (for additional details, see Chap. 5 on Photochemistry).

An example of the variation of the H column abundance in the equatorial region of Saturn with K_h is shown in Fig. 4.1. Voyager 1 and 2 UVS observations yield around 2.5 kR for the equatorial Lyman-α intensity at Saturn, implying an H column abundance above the CH_4 homopause of $\sim 3 \times 10^{16}$ cm^{-2}. From Fig. 4.1, therefore, a value of around 10^8 cm^2 s^{-1} for the eddy diffusion coefficient at Saturn's homopause is obtained. The Lyman-α intensity of Saturn measured with the various instruments is listed in Table 4.1. Note that the Voyager 1 and Voyager 2 Lyman-α data (Table 4.1) include approximately $0.5 \, kR$ contribution due to the interplanetary Lyman-α. The Saturn Lyman-α is thus between 2.5 and 2.8 kR, as assumed above.

The derivation of K_h using the above procedure is strictly valid only under the following assumptions: (1) that resonance scattering alone is responsible for the observed Lyman-α emission, (2) H is produced only from the photochemical processes involving H_2, CH_4, etc., and (3) that the solar Lyman-α flux is known accurately. At Saturn's equatorial and midlatitude regions photochemically produced H is sufficient for explaining the Lyman-α intensity observed during the Voyager encounters and at other times, with resonance scattering being the only excitation mechanism. The situation at Jupiter, however, is not the same. All the Jovian Lyman-α intensities measured to date are listed in Table 4.2. A graphical representation of these, along with the Saturn Lyman-α data is shown in Fig. 4.2. Very little correlation can be seen between the Lyman-α intensities and the solar Lyman-α fluxes, or the $\bar{F}_{10.7}$ cm fluxes, particularly after 1978 (Atreya et al. 1982). The measurements prior to this period have very large statistical uncertainties, and many are done with different instruments, so that calibration errors cannot be disregarded. The above-mentioned lack of correlation implies that either resonance scattering alone is not the sole excitation mechanism, or that H atoms in addition to

Table 4.2. Jupiter Lyman-α

Observation date	Observation technique	Observed Ly-α intensity (kR)[a]	Reference
1967 Dec. 5	Rocket	4.0 $(+8, -2)$	1
1971 Jan. 25	Rocket	4.4 $\pm$ 2.6	2
1972 Sept. 1	Rocket	2.1 $\pm$ 1.0	3
1973 Dec. 3	Pioneer	0.4 $\pm$ 0.12	4
1976 Jan. 5	Copernicus	2.8 $\pm$ 1.0	5
1976 Aug., Sept.	Copernicus	4.0 $\pm$ 1.4	5
1976 Aug., Sept.	Copernicus	3.8 $\pm$ 1.0	6
1978 Mar.	Copernicus	8.3 $\pm$ 2.9	7
1978 Dec. 1	Rocket	13 (central disc)	8
1978 Dec. 10	IUE	12 – 14	9
1979 Mar. – July	Voyager 1 and 2	14 (central disc)	10, 11
1979 May – June	IUE	10 – 13	12
1980 Apr., May	Copernicus	7.0 $\pm$ 2.5	13
1980 May 3	IUE	10	12, 14
1980 July	IUE	10.3	15
1982 Jan.	IUE	9.7	15
1982 Sept.	IUE	10	15
1983 Feb.	IUE	9.8	15
1983 July	IUE	8.5	15

[a] Copernicus intensities have been adjusted for the revised geocoronal calibration according to Bertaux et al. (1980). Pioneer value may be as large as 1 to 1.5 kR following the recalibration of the photometer.
[1] Moos et al. (1969); [2] Rottman et al. (1973); [3] Giles et al. (1976); [4] Carlson and Judge (1974); [5] Bertaux et al. (1980); [6] Atreya et al. (1977a); [7] Cochran and Barker (1979); [8] Clarke et al. (1980b); [9] Lane et al. (1978); [10] Broadfoot et al. (1979); [11] Sandel et al. (1979); [12] Clarke et al. (1980a, 1981b); [13] Atreya et al. (1982); [14] Moos (1981); [15] Skinner (1984).

those produced photochemically must be available. Indeed, approximately 10^{17} H atoms cm^{-2} are needed to explain the Voyager Jupiter data, whereas only 2×10^{16} cm^{-2} are available photochemically (Broadfoot et al. 1981b; Atreya et al. 1982). Additional H atoms can be produced by the charged particle dissociation of H_2 in the auroral region, and then distributed over the entire planet by thermospheric winds (see Chap. 6 on Ionosphere). However, the auroral production rate of the hydrogen atoms is rather uncertain, and the enhancement of the nonauroral H abundances resulting from any thermospheric transport is even more mysterious. Therefore, the above procedure for determining K_h from the planetary Lyman-α intensity is not always applicable.

4.3.2 He 584 Å

The principle behind the determination of K using the Lyman-α data is equally valid for deducing K from resonantly scattered He 584 Å emission from the

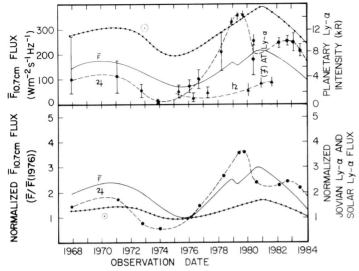

Fig. 4.2. *Top panel* temporal variation of Jupiter ($\mathsf{4}$) and Saturn (h) Lyman-α intensities, $\bar{F}_{10.7\,cm}$ flux ($\bar{F}$) and the solar Lyman-α flux ($\odot$), at Lyman-α in 10^{11} photons cm^{-2} s^{-1}. The actual measured intensities are plotted. *Bottom panel* same as above, except that all parameters are normalized to their respective values on January 5, 1976. Corrections have been applied for contamination by interplanetary Lyman-α, and for conversion from the central disc intensity to the disc average intensity. The calibration correction places the Pioneer 10 intensity at $\sim 1\,kR$ [D.L. Judge 1982, personal communication], as used here. (After Atreya et al. 1982)]

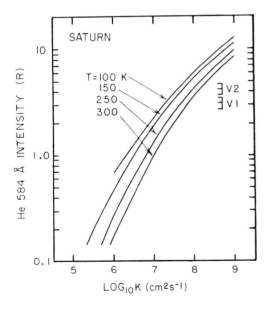

Fig. 4.3. Calculated He 584 Å intensities at Saturn as a function of the homopause eddy diffusion coefficient for various values of the scattering region temperature. The Voyager 1 and 2 (V1 and V2) data and the associated error bars are indicated. (Sandel et al. 1982b)

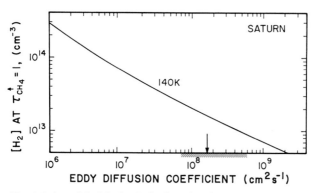

Fig. 4.4. A model of the level of unit optical depth in methane at Lyman-α, $\tau^{\uparrow}_{CH_4} = 1$, expressed in terms of the H_2 density, versus the homopause values of the eddy diffusion coefficient at Saturn. The average temperature in the photochemical regime is ~ 140 K. *Arrow* represents the central values of K_h, whose statistical range is shown by the *stippled bar*. (Atreya 1982)

Jovian planets. Here, increased K would transport more He to the upper atmosphere, resulting in greater He 584 Å emission, unlike the situation with the Lyman-α. However, the relative amount of helium above the scattering layer decreases with increased temperature due to the change in molecular diffusion coefficient, $(D \propto T^S, 1 > s > 0.5)$. Thus, independent knowledge of the temperature in the scattering region, e.g., from the UV stellar occultation data, is essential. The Doppler width of the line must also be known. A model of the variation of Saturnian He 584 Å intensity with K at different temperatures is shown in Fig. 4.3. The Voyager 1 and 2 measurements yield respectively $3.1 \pm 0.4\,R$ and $4.2 \pm 0.4\,R$ for the He 584 Å emission rates at Saturn's disc center (Sandel et al. 1982a; and discussion on calibration in Atreya et al. 1984). The scattering region temperature is around 250 K (Atreya et al. 1984). From Fig. 4.3, then, the range of the eddy diffusion coefficient at the homopause of Saturn is between 7×10^7 and $10^8\,\mathrm{cm^2\,s^{-1}}$. These values are well in agreement with those derived using the Lyman-α data. An important caveat about the procedure employing He 584 Å intensities is that the helium homopause is not necessarily at the same location as the methane homopause. The latter is of greater use in aeronomical applications.

4.3.3 Methane

By far the most direct method for determining K in the H_2-dominated outer planets is by monitoring the distribution of some heavier gas, such as CH_4, whose density is expected to drop off rapidly above the homopause. In principle, once the homopause altitude is known, K_h can be determined, since it is simply equal to the molecular diffusion coefficient at that altitude. The latter is known analytically or empirically for most gaseous mixtures (Mason and

Table 4.3. Eddy diffusion coefficient

	Homopause characteristics				
	Eddy diffusion [$cm^2 s^{-1}$]	Density[a] [cm^{-3}]	Altitude[b] [km]	Pressure [bar]	Reference
Neptune	$10^5 - 10^6$ (?)	$(50-5) \times 10^{13}$	?	$10^{-5} - 10^{-6}$	1
Uranus	$10^4 - 10^6$ (?)	$(100-1) \times 10^{13}$	$350 - 650$	$10^{-5} - 10^{-7}$	2
Saturn	$1.7(+4.3, -1.0) \times 10^8$	1.2×10^{11}	1100	4×10^{-9}	3
	$8.0(+4.0, -4.0) \times 10^7$				4
Jupiter	$1.4(+0.8, -0.7) \times 10^6$	1.4×10^{13}	440	10^{-6}	5
Titan	$1.0(+2.0, -0.7) \times 10^8$	2.7×10^{10}	925 ± 70	6×10^{-10}	6
Earth	$(0.3-1) \times 10^6$	10^{13}	100	3×10^{-7}	7
Venus	10^7	7.5×10^{11}	$130 - 135$	2×10^{-8}	8
Mars	$(1.3-4.4) \times 10^8$	10^{10}	135	2×10^{-10}	9

[a] Density: Atmospheric densities at the homopause correspond to the central values of K_h.
[b] Altitude: For outer planets, the altitudes are above the 1-bar atmospheric pressure level in the equatorial region; some previous publications use the cloud-tops or the 10^{19} cm^{-3} level as the reference. For Titan, Earth, Venus and Mars, the altitudes are above the surface.
[1] See text. [2] Atreya (1984a); Encrenaz et al. (1986); [3] Atreya (1982); [4] Sandel et al. (1982b); [5] Atreya et al. (1981); McConnell et al. (1981); [6] Smith et al. (1982); [7] Hunten (1975); [8] Von Zahn et al. (1980); [9] Nier and McElroy (1977).

Marrero 1970). Photochemical processes, however, can result in the depletion of CH_4 well below the homopause, giving a false impression of homopause. Hence, determination of K using the CH_4 data is based on comparing photochemical models of CH_4 profiles with the measurements (Atreya et al. 1981; Atreya 1982). The photochemical models are highly sensitive to the choice of the eddy diffusion coefficient. An example of such calculations for Saturn is shown in Fig. 4.4. In this figure, the level of the unit optical depth in methane at Lyman-α ($\tau_{CH_4} = 1$) − given in terms of the H_2 density − is plotted as a function of K_h. The Voyager δ Sco stellar occultation data give this level to be where the H_2 density is between 9×10^{12} and 2.5×10^{13} cm^{-3}, (Festou and Atreya 1982). The above values of H_2 at the $\tau_{CH_4} = 1$ level would imply a range of 7×10^7 to 6×10^8 $cm^2 s^{-1}$ for K_h, using Fig. 4.4. Again, the central value of K_h is similar to the one obtained from the other two procedures. At Jupiter, however, such agreement is lacking due to previously mentioned complications. Current understanding of planetary eddy diffusion coefficients at the homopause is summarized in Table 4.3. Note that at Saturn, where K_h is large, the $\tau_{CH_4} = 1$ level is nearly 300 km below the homopause since the solar photons deplete CH_4 already much below the homopause. With lower values of K, e.g., at Jupiter, the discrepancy between these two altitudes is smaller.

4.3.4 Eddy Diffusion on Uranus and Neptune

The intensity of Lyman-α emission is less of an indicator of the atmospheric vertical mixing on Uranus than on Jupiter and Saturn. This is due to the fact that the observed Lyman-α emission from Uranus is composed of three components – resonantly scattered solar Lyman-α, backscattered interplanetary/interstellar medium Lyman-α, and the "electroglow". Electroglow refers to the emission from the "sunlit" hemisphere due to soft electron excitation. For Lyman-α, both direct excitation by soft electrons as well as solar resonance scattering by hydrogen atoms produced on dissociation of H_2 by the soft electrons are included. In the context of Uranus, the energy distribution of these electrons is characterized by a low electron temperature of 3 eV.

The term electroglow was introduced following the Voyager/UVS observations of Uranus (Broadfoot et al. 1986), as the emission characteristics did not fit conventional definitions of dayglow or the aurora. Dayglow results from the action of solar EUV (photons and photoelectrons), whereas aurora is the result of precipitation of energetic charged particles in the "auroral zone". Although the electroglow is confined to the sunlit hemisphere of Uranus, it is not caused by the solar EUV, as the power input implied by the observed hydrogen intensities, 2×10^{11} W, is a factor of 100 greater than that available from the solar EUV at Uranus. Furthermore, electroglow contains emissions of both H Lyman-α and the H_2 Lyman and Werner bands; the latter are excited by electrons. Electroglow is not strictly an auroral emission either, as it is present over the entire sunlit hemisphere of Uranus, not just in the auroral zone. Furthermore, electroglow is caused by soft electrons, rather than the energetic particles which are responsible for an aurora. To be sure, auroral excitation does indeed occur on Uranus. It was detected by Voyager/UVS in the vicinity of the south magnetic pole which at that time was located in the dark (northern) hemisphere. There is no electroglow on the dark side of Uranus. The source and energization mechanism of the electrons responsible for electroglow are as yet unknown. Until the extent of contribution of the various above-mentioned sources of Lyman-α from Uranus are determined, some tracer other than Lyman-α is needed to measure K on Uranus.

The analysis of γ Pegasi and ν Geminorum ultraviolet stellar occultation monitored by Voyager provides the best hope for determining the eddy diffusion coefficient at the homopause of Uranus. These data are expected to provide the CH_4 density distribution in the upper atmosphere. A comparison of the measured CH_4 density profile with the photochemical models will fix the value of K, as has been done successfully for Jupiter and Saturn.

With relatively small solar energy input at Uranus, combined with its weak internal heat source, one would not expect as vigorous a vertical mixing on Uranus as on Saturn. A nominal K of $10^5 - 10^6 \, \text{cm}^2 \, \text{s}^{-1}$ at the homopause of Uranus is suggested by Atreya (1984a). The CH_4 photochemical models done by Atreya (1984a) yield the vapor-phase C_2H_2 column abundance (after accounting for saturation), which, when compared with the IUE data on C_2H_2

between 1600 and 1850 Å, indicates a value of about $10^5 \, cm^2 \, s^{-1}$ for the eddy diffusion coefficient at the Uranus homopause (Encrenaz et al. 1986). Similar values of K are suspected for Neptune also. Although Neptune receives even less solar energy than Uranus, it does, however, possess a strong internal heat source.

4.4 Eddy Diffusion in the Homosphere

The behavior of K below the homopause is a function of the atmospheric number density (n) and the thermal structure. From considerations of incoherent wave motion, which would cause vertical transport, and the diurnal tide in the Earth's atmosphere, Lindzen (1971) concluded that K varies roughly as the inverse square root of the atmospheric number density (strictly speaking, such a behavior is valid only if wave energy and static stability are height-independent). Since the thermal structure in the middle atmospheres of the major planets is unknown, the precise behavior of eddy diffusion coefficient in this region cannot be predicted with confidence. In photochemical calculations, however, K is assumed to vary with n alone. The assumption works well, except perhaps near the tropopause. Rapid and sudden change in the value of K in the terrestrial troposphere, particularly near the tropopause, is well known (Hunten 1975). Theoretical interpretation of the stellar occultation data for the hydrocarbons indicates that for all practical purposes, the eddy diffusion coefficient varies inversely as the square root of the atmospheric number density (i.e., $K \propto n^{-k}$, $k = 0.5$) at Jupiter and Saturn (Atreya et al. 1981; Atreya 1982). The actual value of k was found to lie between 0.5 and 0.6. The tropopause (~ 100-mb, level) values of the eddy diffusion coefficient extrapolated from those at the homopause are $2 \times 10^3 \, cm^2 \, s^{-1}$ and $1.2 \times 10^4 \, cm^2 \, s^{-1}$, respectively, at Jupiter and Saturn. Again, extreme caution should be exercised in employing such an extrapolation since the thermal structure of the intermediate region (through which the above extrapolation occurred) is yet unknown.

4.5 Molecular Diffusion Coefficient

Even though eddy diffusion dominates in the homosphere, the molecular diffusion coefficient would be greater than K in the upper atmosphere as it grows faster with decreasing atmospheric density. The molecular diffusion coefficient can be estimated from the classical gas kinetic theory considerations. The following general expression for D_{12}, the coefficient for diffusion of gas 1 through gas 2, is obtained (Chapman and Cowling 1970).

$$D_{12} = \frac{3\pi}{32 \bar{Q}_D n} v_{12} \, , \tag{4.19}$$

Table 4.4. Molecular diffusion coefficients relevant to the outer planets, $(D_{12} = A T^S/n)$ [a]

Binary System	A $[\times 10^{17}]$	s	$n D_{12} = A T^S$ $[\times 10^{19} \, cm^{-1} s^{-1}]$
$H - H_2$	8.19	0.728	3.44
$H - He$	10.30	0.732	4.42
$H - N_2$	4.87	0.698	1.39
$H_2 - N_2$	2.80	0.740	1.25
$CH_4 - H_2$	2.27	0.765	1.15
$CH_4 - N_2$	0.725	0.750	0.34
$Ar - N_2$	0.655	0.752	0.031

[a] A and s are taken from Mason and Marrero (1970). T is assumed 170 K. In addition to the binary systems listed in the above table, C_2H_2, C_2H_6, and C_2H_4 diffusing through H_2 at 170 K (a typical Jovian middle atmospheric situation), all have $n D_{12} \approx 0.9 \times 10^{19} \, cm^{-1} s^{-1}$ (i.e., $A \approx 6.86 \times 10^{17}$ and $s = 0.5$).

where, $\bar{Q}_D$ is the mean momentum transfer cross-section, $n = n_1 + n_2$ is the total concentration, and v_{12} is the velocity of gas 1 through gas 2; it is further given by

$$v_{12} = (8 kT/\pi \mu)^{1/2} , \tag{4.20}$$

where,

$$\mu = (1/m_1 + 1/m_2)^{-1} \tag{4.21}$$

is the reduced mass in g.

For a minor constituent diffusing through the background atmosphere, n becomes the atmospheric density ($n \approx n_2$), and m_2 the mean molecular weight. Exact theories and experimental data reduce the abovementioned expression of D to the following form (Mason and Marrero 1970; Banks and Kockarts 1973),

$$D_i = b_i/n = A T^S/n , \tag{4.22}$$

where b_i is called the binary collision parameter. It is expressed in terms of coefficients A and s. The values of A and s for various gases of relevance to the outer planets are given in Table 4.4.

Since D_{12} grows as n^{-1}, while K grows as $n^{-0.5}$, D_{12} would surpass K at some height in the atmosphere. This altitude is, of course, the homopause level. The variations of K and D with the atmospheric density are shown for Jupiter in Fig. 4.5. The data are taken from Table 4.3 (extrapolated with $K \propto n^{-0.5}$) and Table 4.4 ($CH_4 - H_2$ system). As expected, the $K - D$ crossover occurs at the homopause level given in Table 4.3.

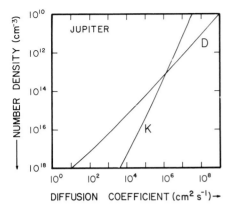

Fig. 4.5. Behavior of eddy (K) and molecular (D) diffusion coefficients with the atmospheric number density on Jupiter

4.6 Outstanding Issues

What is the behavior of the eddy diffusion coefficient below the homopause, i.e., through the middle atmosphere, across the tropopause and in the upper and lower troposphere? How does the vertical mixing vary from equator to the pole, and also temporally? Is there a verifiable effect of internal heat on the tropospheric vertical mixing? Why is Saturn's eddy mixing coefficient so much greater than Jupiter's at the homopause? Does the discrepancy propagate all the way down to the troposphere? What is the origin of electroglow electrons, and how do they receive their power? What are the relative contributions of electroglow, resonance scattering and the interstellar medium to the planetary Lyman-α intensity? These issues are intimately related to the planetary thermal structure. Refer to Chapter 2.6 on *Thermal Structure* for resolutions which are equally applicable here as there.

5 Photochemistry

The first published account of the importance of photochemistry in the Jovian atmosphere is that of Wildt (1937), who surmised the possibilities for the recycling of CH_4 and NH_3 following their photodecomposition. Numerous minor constituents are produced in a planetary atmosphere, predominantly by photochemical processes, subsequent to the absorption of solar ultraviolet photons by various atmospheric molecules. The knowledge of the distribution of minor constituents can be important in answering many key questions concerning atmospheric transport, meteorology, atmospheric stability, physics of the interior, and atmospheric coupling to the exterior. On all the major planets, CH_4 undergoes photochemical transformation. On Uranus and Neptune, it is expected to form a cloud as well, whereas its products, C_2H_2, C_2H_6, and polyacetylenes, are likely to condense in the form of tropospheric/stratospheric hazes. On Jupiter and Saturn, NH_3 forms a visible cloud. Its saturated vapor abundance above the clouds, however, is large enough for it to be photolyzed as well. For all practical purposes, the NH_3 photochemistry is both spectroscopically and physically separated from the CH_4 photochemistry. This is because CH_4 is photolyzed below 1600 Å (mostly by the Lyman-α photons which have a large flux), and NH_3 between 1600 Å and 2300 Å. Furthermore, gaseous NH_3 abundance on Jupiter and Saturn is negligible above atmospheric pressure levels of ~ 0.1 bar. In a small altitude region, coupling between the photochemical products of NH_3 and CH_4 may occur, resulting in the formation of, perhaps, methylamine (CH_3NH_2), hydrogen cyanide (HCN), etc. Other interesting photochemical species are H_2S and PH_3, both of which have been proposed as potential candidates for the formation of chromophores needed to explain the colors of Jovian clouds. The presence of carbon monoxide (CO), germane (GeH_4) and phosphine (PH_3) in the stratosphere is indicative of strong convection in the interior. In all photochemical processes on the outer planets, atomic hydrogen, which is itself produced on the dissociation of H_2, is the most important intermediate reactant. The present status of the photochemistry of NH_3, PH_3, CH_4, H_2S, CH_3NH_2, HCN, and CO on the outer planets is discussed below.

5.1 Hydrogen

The production and loss of atomic hydrogen is largely controlled by the ionospheric processes, as will be discussed in detail in Chapter 6 on the Iono-

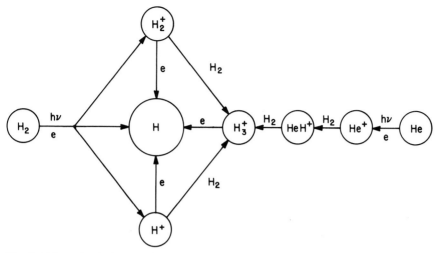

Fig. 5.1. Photochemical scheme for hydrogen production on the major planets. Both solar photons as well as charged particles can dissociate and ionize H_2

sphere. At low and middle latitudes atomic hydrogen is produced by photodissociation, dissociative photoionization, and from ionization of H_2. The dissociation and ionization continua of H_2 lie below 845 and 805 Å respectively. The major pathways for the H production are:

$$H_2 + h\nu \rightarrow H + H$$
$$\rightarrow H_2^+ + e^-$$
$$\rightarrow H^+ + H + e^-$$

followed by

$$H_2^+ + H_2 \rightarrow H_3^+ + H$$

and

$$H_3^+ + e^- \rightarrow H_2 + H$$
$$\rightarrow H + H + H$$

or

$$H^+ + e^- \rightarrow H + h\nu.$$

A small ($\sim 20\%$) contribution to the H_3^+ (hence H) concentration is provided by reactions between He^+ and H_2. The complete chemical scheme for the ionospheric source of H is shown in Fig. 5.1. A small fraction of H also results from fluorescent dissociation of H_2 following discrete absorption in the Lyman and Werner bands. The atomic hydrogen produced in the ionosphere is transported to the dense atmosphere where its major loss is via 3-body recombination. In the lower atmosphere, the photolysis of CH_4 and NH_3 is also a source of atomic hydrogen. At high latitudes, hydrogen production is dominated by the charged particle dissociation of H_2. Additional details of the hydrogen photochemistry are found in Chapter 6 on Ionosphere.

5.2 Ammonia (NH₃)

The white regions seen in the visible images of Jupiter and Saturn (e.g., Figs. 2.22 and 2.23) represent the tops of ammonia clouds. These are generally the areas of "zones", where rising air motion results in condensation of NH_3 at cooler temperatures. The Voyager infrared, visible, and radio observations place the cloud tops at around the 0.6-bar level on Jupiter and somewhat deeper on Saturn. The orange-brown hue seen in some of the clouds, most prominent of which is the Great Red Spot (GRS) on Jupiter, is perhaps caused by the presence of some chromophore (see later PH_3, H_2S, and HCN). The gas-phase abundance of NH_3 above the clouds is dictated by its saturation vapor pressure. The level of penetration of the UV photons is determined, not only by the physical barrier provided by the clouds, but also the Rayleigh scattering optical depth. For a molecular atmosphere (such as H_2 and He), the Rayleigh scattering cross-section per particle is given by the following expression (Chandrasekhar 1960)

$$\sigma_R = \frac{8\,\pi}{3} \left(\frac{2\,\pi}{\lambda} \right)^4 \alpha^2, \tag{5.1}$$

where λ is the wavelength (in cm) of radiation being scattered by the molecules whose polarizability is α. The polarizability of H_2 and He are, respectively, 0.82×10^{-24} and 0.21×10^{-24}. With these values of α, and λ expressed in units of 10^{-3} Å, the expression of Rayleigh scattering cross-section (in cm^2) in the atmosphere of Jupiter takes the following form.

$$\sigma_R = 7.5 \times 10^{-25} / \lambda^4 \tag{5.2}$$

(λ in 10^{-3} Å).

The above expression for σ_R gives unit optical depth due to Rayleigh scattering on Jupiter at the atmospheric column density levels of 10^{26} cm^{-2} (~ 1 bar), 2×10^{25} cm^{-2} (~ 200 mb), and 5×10^{17} cm^{-2} (~ 10 mb), respectively, for solar photons of wavelength 3000, 2000, and 1000 Å. This means that Rayleigh scattering would inhibit solar photons shortward of 2000 Å from penetrating below the 10^{19} cm^{-3} (200 mb) level. The absorption of solar photons by NH_3 occurs mainly between 1600 and 2300 Å (see Fig. 5.2 for NH_3, N_2H_4, and PH_3 absorption cross-sections) on Jupiter and Saturn. On Uranus and Neptune, NH_3 is buried well below the CH_4-ice clouds (see Chap. 3 on Cloud Structure), therefore its photolysis is improbable. The photochemical scheme for NH_3 at Jupiter and Saturn is presented in Table 5.1, and schematically shown in Fig. 5.3. The chemical scheme, taken from Atreya et al. (1977b),

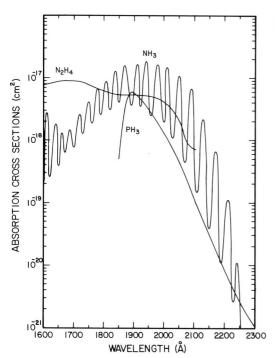

Fig. 5.2. Photoabsorption cross-sections of NH$_3$, PH$_3$, and N$_2$H$_4$

Table 5.1. Ammonia photochemistry[a]

Chemical reactions	Rate constant[b]	Reaction number	Reference
NH$_3$ + $h\nu \rightarrow$ NH$_2$($\tilde{X}^2$B$_1$)	J(NH$_3$)	R1	1
NH$_2$(X) + H + M $\rightarrow$ NH$_3$ + M	$k_2 = 6 \times 10^{-30}$[M]/(1 + 3 × 10^{-20}[M])	R2	2
NH$_2$(X) + NH$_2$(X) + M $\rightarrow$ N$_2$H$_4$ + M	$k_3 = 1 \times 10^{-10}$	R3	2
N$_2$H$_4$ + H $\rightarrow$ N$_2$H$_3$ + H$_2$	$k_4 = 9.87 \times 10^{-12}$ exp($-1198/T$)	R4	3
N$_2$H$_4$ + $h\nu \rightarrow$ N$_2$H$_3$ + H	J(N$_2$H$_4$)	R5	1
N$_2$H$_3$ + H $\rightarrow$ 2 NH$_2$	$k_6 = 2.7 \times 10^{-12}$	R6	4
N$_2$H$_3$ + N$_2$H$_3$ $\rightarrow$ 2 NH$_3$ + N$_2$	$k_7 \ll k_8$	R7	5
N$_2$H$_3$ + N$_2$H$_3$ $\rightarrow$ N$_2$H$_4$ + N$_2$ + H$_2$ $\rightarrow$	$k_8 = 6 \times 10^{-11}$	R8a	5
N$_2$H$_4$ + N$_2$H$_2$		R8b	
H + H + M $\rightarrow$ H$_2$ + M	$k_9 = 8 \times 10^{-33}$(300/T)$^{0.6}$	R9	6

[a] After Atreya et al. (1977b).

[b] Rate constants in units of cm^3 s^{-1} for 2-body and cm^6 s^{-1} for 3-body reactions, photodissociation rate J is in s^{-1}.

[1] see text; [2] Gorden et al. (1971); [3] Stief and Payne (1976); [4] Gehring et al. (1971); [5] L. J. Stief (1976, pers. comm.); [6] Ham et al. (1970).

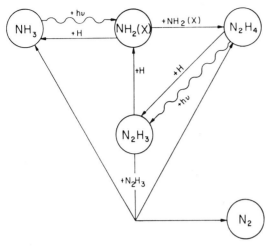

Fig. 5.3. Photochemical scheme for ammonia on the major planets. (Atreya et al. 1978)

incorporates the latest available laboratory and atmospheric data, as well as hydrazine (N_2H_4) saturation and the subsequent chemistry. Initial photochemical calculations for NH_3 were done by Strobel (1973a). The photolysis rates for the gaseous NH_3 and N_2H_4 at Jupiter approach a value, J_∞, of 3.0×10^{-6} s^{-1} and 1.3×10^{-6} s^{-1}, respectively, at an altitude of approximately 35 km above the 1-bar level.

The photolysis rate of the ith constituent at wavelength λ is given by

$$J_i(z, \lambda) = \sigma_a^i(\lambda) F(z, \lambda) , \tag{5.3}$$

where $\sigma_a^i(\lambda)$ is the absorption cross-section of the ith constituent at wavelength λ, $F(z, \lambda)$ is the solar flux at wavelength λ and altitude z.

$F(z, \lambda)$ can be further expressed as

$$F(z, \lambda) = F_\infty(\lambda) \exp[-\tau(z, \lambda)] , \tag{5.4}$$

where $\tau(z, \lambda)$ is the atmospheric optical depth at wavelength λ, so that for a multicomponent atmosphere.

$$\tau(z,\lambda) = \sum_i \sigma_a^i(\lambda) \int_z^\infty n_i(z) dz \sec\chi , \tag{5.5}$$

where $ds = dz \sec \chi$ \hfill (5.6)

is the slant path, and χ is the solar zenith angle.

The total photolysis rate of the ith constituent at altitude z is then given by the following expression.

$$J_i(z) = \sum_\lambda J_i(z,\lambda) . \tag{5.7}$$

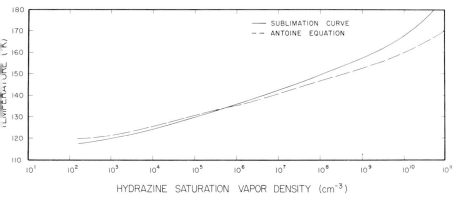

Fig. 5.4. N$_2$H$_4$ saturation vapor density (in particles cm^{-3}) from considerations of the latent heat for sublimation. The value at 100 K is 5.6 N$_2$H$_4$ particles cm^{-3}. (After Atreya et al. 1977b)

The dissociation of NH$_3$ produces amidogen radicals, NH$_2$(X), in the ground state (R1, Table 5.1). NH($a^1\Delta$) may also be produced, however its quantum yield is less than 4%; moreover it also reacts quickly with H$_2$ to form NH$_2$ (Schurath et al. 1969; McNesby et al. 1962). The reaction of NH$_2$(X) with H recycles up to one third of NH$_3$. The self-reaction of NH$_2$(X), in the presence of a background gas (denoted by M in Table 5.1), however, results in the formation of N$_2$H$_4$ (R3). A large fraction of N$_2$H$_4$ is susceptible to condensation in the Jovian and Saturnian tropospheres because of its relatively low vapor pressure. This becomes quite apparent from the sublimation curves of N$_2$H$_4$ shown in Fig. 5.4. N$_2$H$_4$ condensate could most likely be the "Axel dust" postulated by Axel (1972)[1]. Both with and without supersaturation, the N$_2$H$_4$ present in the gas phase would undergo photolysis (R5), and the hydrogen extraction reaction (R4). Condensation of hydrazyl (N$_2$H$_3$) also occurs quite readily at the Jovian and Saturnian tropospheric temperatures. After several intermediate reactions involving N$_2$H$_3$ vapor, R6 – R8, N$_2$ would be formed. Possible formation of diimide (N$_2$H$_2$) in reaction R8b can reduce the yield of N$_2$. Direct formation of N$_2$, R8a, however is the preferred pathway.

The distribution of the various constituents of NH$_3$ photochemistry is calculated by solving one-dimensional coupled continuity and eddy transport equations (Chap. 4 on Vertical Mixing). These equations form a set of coupled nonlinear partial differential equations which can be solved by the standard numerical techniques, such as the generalized Newton-Raphson's method (see Ames 1969). The rate of change of NH$_3$, using the chemistry in Table 5.1, for example, is given by

$$\frac{\partial}{\partial t}[NH_3] = -[NH_3]\,J(NH_3) + k_2[NH_2][H] + 2\,k_7[N_2H_3]^2,$$

---

[1]   In the literature, atmospheric aerosols, including hydrocarbon hazes, are also referred to as Danielson dust, in memory of R. E. Danielson.

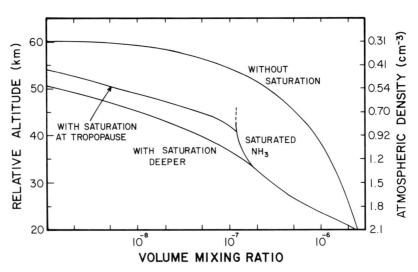

Fig. 5.5. Photochemical distribution of ammonia at the midlatitudes of Jupiter under various assumptions. Altitudes are above the 1-bar level. Atmospheric densities corresponding to the altitudes are indicated on the right ordinate. The curve marked "without saturation" does not limit the photochemical distribution to the value allowed by saturation as the calculations proceed higher in the troposphere. This is clearly an unrealistic situation. The curve *saturated NH₃* is simply NH₃-saturated vapor mixing ratio, it continues upward as indicated by the *broken line*. The curve *with saturation at tropopause* is the photochemical profile obtained by forcing the calculated NH₃ to the saturated value at the tropopause (below the tropopause, calculated NH₃ does not exceed the saturated value). Without large amounts of stratospheric/tropospheric haze, the UV photons could penetrate to somewhat deeper levels (limited by Rayleigh scattering). In that case, the distribution would be as shown by the curve, *with saturation deeper*. (Atreya and Romani 1985)

which is equal to zero in steady state. The square brackets represent number densities. A typical midlatitude NH_3 profile appropriate for the Voyager/Jupiter encounter is shown in Fig. 5.5. Note in this figure that without saturation, NH_3 abundances are much greater after photolysis than in the case where NH_3 is forced back to its saturation value wherever the NH_3 partial pressure exceeds the saturation vapor pressure. Above the tropopause, however, photochemistry, and transport deplete NH_3 to well below its saturation limit. The actual photochemical distribution depends on the depth of penetration of the solar photons, which in turn can be influenced by the extent of atmospheric haze and clouds. In the equatorial region, the NH_3 abundances would be somewhat lower than those shown in Fig. 5.5 due to the greater photolysis rate; the opposite would hold for the high latitudes. The Voyager IR measurements (Kunde et al. 1982) yield NH_3 mixing ratios of 2×10^{-7} at 0.2 bar $(1.2 \times 10^{19} \, cm^{-3})$ and 1.5×10^{-5} at 0.5 bar $(2.5 \times 10^{19} \, cm^{-3})$ which are in good agreement with the model of Fig. 5.5. The calculated distribution of the photochemical products of NH_3 on Jupiter is shown in Fig. 5.6. A comparison between the hydrazine distributions with and without supersaturation indi-

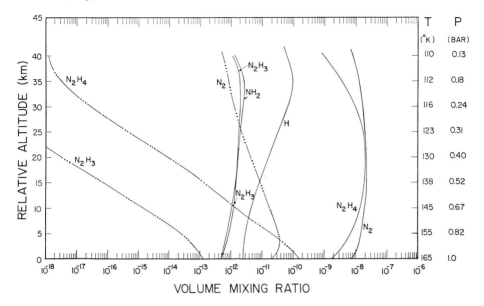

Fig. 5.6. Distributions of the photochemical products of NH$_3$. Pressures and temperatures corresponding to the altitudes are shown on the right ordinate. N$_2$ distributions are shown for the case where N$_2$H$_4$ is enormously supersaturated (*solid line*), and where N$_2$H$_4$ mixing ratio is limited to its saturated value ($-\cdots-$). (After Atreya et al. 1977b)

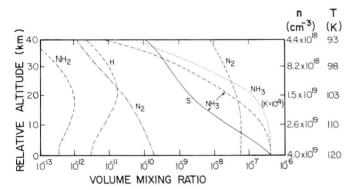

Fig. 5.7. Same as Fig. 5.6, except for Saturn. Curve S represents the saturated volume mixing ratio of NH$_3$. The larger of the two N$_2$ distributions refers to the case with the N$_2$H$_4$ supersaturation, the other is for the case with saturated N$_2$H$_4$. (Atreya et al. 1980)

cates that virtually all of the hydrazine above the cloud tops is susceptible to condensation. The maximum N$_2$H$_4$ column ice production rate is calculated to be 7×10^{10} molecules cm^{-2} s^{-1}, or approximately 1 mg m^{-2} per Jovian day (Atreya et al. 1977b). The range of N$_2$ mixing ratios, 10^{-8} to 4×10^{-11}, results from the two extremes of N$_2$H$_4$ with and without supersaturation (Atreya et al. 1977b). On the other hand, Prinn and Olaguer (1981) have suggested that vertical motions on Jupiter may be large enough to transport large amounts of N_2 from the interior to the upper atmosphere. They predict N_2 volume mixing ratios of $(0.6 - 3) \times 10^{-6}$ if iron-catalyzed reaction between N_2 and (3)H_2 is important in the deep atmosphere, and 10^{-5} if it is not. A test of whether N_2 is produced photochemically (Atreya et al. 1977b), or is upwelled from the interior (Prinn and Olaguer 1981) will have to wait until after the data from the Galileo Probe Mass Spectrometer have been obtained, analyzed, and interpreted. Photochemical calculations for NH_3 on Saturn done in the same manner as for Jupiter are shown in Fig. 5.7. Again, there is a distinct possibility of N_2H_4 haze formation. Voyager IR data for the tropospheric ammonia are sparse. They do, however, tend to favor the saturated NH_3 profile of the model shown in Fig. 5.7.

5.3 Phosphine (PH_3)

PH_3 is thermochemically stable below the 800 K level in the deep atmosphere of Jupiter. Between 800 and 300 K, PH_3 can be oxidized to P_4O_6 by H_2O. Above the 300 K level, P_4O_6 would dissolve in H_2O. However, PH_3 has been detected in the troposphere and stratosphere of Jupiter and Saturn. Prinn and Lewis (1975) have argued that strong vertical mixing can transport PH_3 faster to the upper atmosphere than the time constant for the oxidation of PH_3 to P_4O_6. The latter is estimated to be 6 days, with $K = 10^9 \, cm^2 \, s^{-1}$ (Prinn and Owen 1976). Once above the ammonia cloud tops, PH_3 undergoes photolysis at and above the 80-mb level following absorption of the solar photons between 1600 and 2350 Å. The photolysis of PH_3 thus takes place in essentially the same wavelength range as of NH_3. However, since the PH_3 abundance at all altitudes of interest is lower than the NH_3 abundance on Jupiter, the attenuation of solar flux by PH_3 is quite small compared to that of NH_3. The chemical scheme for the PH_3 photolysis is shown in Table 5.2. Unlike NH_3, phosphine is not recycled by the $(H + PH_2)$ reaction according to this scheme. The final product is triclinic red phosphorus. (In the laboratory experiments of Ferris and Benson (1980), phosphorus with colors ranging from yellow to violet is produced.) These phosphorus crystals can provide the reddish coloration seen in the GRS and the other clouds, provided that the density of $P_4(s)$ is large enough for about unit optical depth in the visible. Vertical eddy diffusion coefficient, K, of $10^6 \, cm^2 \, s^{-1}$ throughout the upper troposphere and the interior would be required for $\tau_{P_4} = 1$ (Prinn and Lewis 1975). $K \approx 10^6 \, cm^2 \, s^{-1}$ has been measured only near the homopause (Atreya et al. 1981), which implies a relatively low value of 5×10^2 to $10^3 \, cm^2 \, s^{-1}$ for K near the tropopause. This value of K, however, is valid for the quiescent region, and could indeed be rather large for the Great Red Spot region. Furthermore, tropospheric eddy diffusion could have an unpredictable behavior, as on the Earth (Hunten

Phosphine (PH_3)   89

Table 5.2. Photochemistry of PH_3[a]

Chemical reactions	Reaction number
$PH_3 + h\nu \rightarrow PH_2 + H$	R1
$PH_2 + PH_2 \rightarrow PH + PH_3$	R2
$PH_3 + H \rightarrow PH_2 + H_2$	R3
$PH + PH \rightarrow P_2 + H_2$	R4
$P_2 + P_2 \rightarrow P_4(g)$	R5
$P_4(g) \rightarrow P_4(s)$	R6
$H + H + M \rightarrow H_2 + M$	R7

Reaction (R3) may compete with reaction (R7) at high temperatures.
[a] After Prinn and Lewis (1975).

1975). In any event, the value of K in the interior is expected to be greater than that near the tropopause.

Two important considerations, however, could dilute the above arguments for the formation of red phosphorus crystals, (a) scavenging of intermediate products by the hydrocarbons, and (b) possible formation of diphosphine (P_2H_4) which could prematurely terminate the chemistry.

The laboratory simulation experiments of Vera-Ruiz and Rowland (1978) indicate that C_2H_2 and C_2H_4 act as efficient scavengers of PH, P_2 and PH_2, thus suppressing the formation of the $P_4(s)$ crystals. After Prinn and Lewis, they have suggested that strong vertical mixing could transport PH_3 to the upper atmosphere, where it would be photolyzed, and thus prevent the scavenging action of C_2H_2 and C_2H_4. The Voyager IR measurements (Kunde et al. 1982), however, yield fairly high abundances of C_2H_2 and C_2H_4 (which are consistent with the proportions taken in the experiment of Vera-Ruiz and Rowland) precisely in the region where the photolysis of PH_3 occurs. The formation of red phosphorus crystals is thus questionable. An important caveat to consider in regard to the measurements of Vera-Ruiz and Rowland is that no information on the activation energies at the relevant Jovian temperatures (150 – 110 K) is available for reactions (R2), (R4) and (R5) of Table 5.2, or for the scavenging reactions which compete with these reactions. An additional difficulty in the formation of $P_4(s)$ may arise due to possible termination of the PH_3 photochemistry subsequent to the formation of diphosphine, P_2H_4, by the self-reaction of PH_2. Ferris and Benson (1980) have found in their laboratory experiments that absorption of 2062 Å photons by PH_3 resulted in the production of P_2H_4. The yield of P_2H_4 rose to a maximum in 5 h of illumination time and then dropped to 20% of the maximum. The proposed chemical scheme for PH_3 photolysis, which differs from the earlier ones in the above-mentioned sense, is given in Table 5.3. It is rather similar to the one for NH_3 (Table 5.1), i.e., following the photolysis of PH_3, PH_2 radicals react with themselves to form P_2H_4, similar to the formation of N_2H_4 in the NH_3-chemistry. No reliable thermodynamical data on P_2H_4 are

Table 5.3. Photochemistry of PH_3[a]

Chemical reactions	Reaction number
$PH_3 + h\nu \rightarrow PH_2 + H$	R1
$PH_2 + PH_2 + M \rightarrow P_2H_4 + M$	R2
$P_2H_4 + H \rightarrow P_2H_3 + H_2$	R3
$P_2H_4 + PH_2 \rightarrow P_2H_3 + PH_3$	R4
$P_2H_3 + H \rightarrow P_2H_2 + H_2$	R5
$P_2H_3 + P_2H_3 \rightarrow P_2H_2 + P_2H_4$	R6
$P_2H_3 + PH_2 \rightarrow P_2H_2 + PH_3$	R7
$P_2H_2 + h\nu \rightarrow P_2 + H_2$	R8
$P_2 + P_2 \rightarrow P_4(g) \rightarrow P_4(s)$	R9
$H + H + M \rightarrow H_2 + M$	R10

[a] After Ferris and Benson (1980).

available. However, it is suggested that P_2H_4 might behave similarly to N_2H_4, and may undergo rapid condensation in the Jovian stratosphere. This can choke off the production of large quantities of P_4. As with most laboratory kinetics work, the measurements of Ferris and Benson were done at room temperature and pressure, and no excess of H_2 was allowed, i.e., Jovian conditions were not simulated. The real potential of the phosphorus chromophore hypothesis cannot be evaluated until relevant laboratory data and the measurements of Jovian tropospheric dynamics become available. Finally, note that there is something puzzling about the PH_3 mixing ratios on Jupiter and Saturn determined by the Voyager IR spectrometer. The value in the lower atmosphere of Jupiter is found to be close to that expected for a solar composition atmosphere; it is, however, about two times greater at Saturn (although it should be recognized that the error bars in the PH_3 mole fraction at Saturn are large enough to accommodate even a solar value). Even the C/H ratio on Saturn is much larger than solar (about five times). If these ratios are indeed indicative of their bulk abundances (i.e., interior plus atmosphere), they might imply strong inhomogeneities in the solar nebula from which these planets presumably accreted.

5.4 $NH_3 - PH_3$ Coupling and NH_3 Recycling

Some possibility exists of the reaction between NH_2, which is produced on photodissociation of NH_3, and PH_3 (Strobel 1977). This would be an important channel for recycling NH_3 at Jupiter. In particular, reactions listed in Table 5.4 are involved. The above mechanism would be operative provided that the reaction of NH_2 with PH_3 (R3, Table 5.4) is faster than its self-reaction (R2), and the conversion of PH_3 to PH_2 (R4). In fact, laboratory measurements done at the appropriate Jovian temperature (150 K) show that the rate for R3 is 3.2×10^{-15} $cm^3 s^{-1}$, which is a factor of 10,000 slower than the rate

Table 5.4. NH$_3$ – PH$_3$ coupling

Chemical scheme	Reaction number
NH$_3$ + $h\nu$ → NH$_2$ + H	R1
NH$_2$ + NH$_2$ + M → N$_2$H$_4$ + M	R2
NH$_2$ + PH$_3$ → NH$_3$ + PH$_2$	R3
PH$_3$ + H → PH$_2$ + H$_2$	R4

for R$_2$, and a factor of 100 slower than the rate for R4. Hence, it is clear that there cannot be any significant coupling between the photochemistries of PH$_3$ and NH$_3$, and that some other mechanism must be responsible for recycling ammonia at Jupiter. Taking into consideration transport and convection, all of the ammonia above and below the clouds of Jupiter should have been irreversibly converted by the action of solar photons to other forms in 60 million years. The present-day abundance of NH$_3$ on Jupiter, on the contrary, is about the same as that of a solar composition atmosphere. Several scenarios are plausible for recycling NH$_3$ (Strobel 1973a; Atreya and Donahue 1979). These involve mixing of N$_2$H$_4$ or N$_2$ to the deep interior. The hydrazine produced in the NH$_3$ photochemistry would eventually snow out of the atmosphere and mix to the deep hot interior of Jupiter, where it would be thermally dissociated to NH$_2$, i.e.,

$$N_2H_4 \xrightarrow{\text{high temp.}} NH_2 + NH_2$$

followed by

$$2\,NH_2 + H_2 \rightarrow 2\,NH_3 .$$

The last reaction has a large activation energy, therefore it cannot recycle NH$_3$ in the cold troposphere of Jupiter. Another possibility is that N$_2$ produced in the NH$_3$ photochemistry convects to the deep hot interior of Jupiter, where at a pressure of $\sim$3000 bar, it would recombine with H$_2$ in the following ter-molecular reaction,

$$N_2 + 3\,H_2 \rightarrow 2\,NH_3 .$$

NH$_3$ so produced would mix to the upper troposphere. Again, this reaction cannot proceed in the cold troposphere of Jupiter, and is possible only after N$_2$ convects to the interior. Similar ideas are applicable to Saturn; however, photolysis rates there are much slower.

5.5 Hydrogen Sulfide (H$_2$S)

Hydrogen sulfide has not been detected on the major planets. Nonetheless, Lewis and Prinn (1970) proposed that the colors of the Jovian clouds could be

Table 5.5. Photochemistry of H_2S [a]

Chemical reactions	Reaction number
$H_2S + h\nu \rightarrow HS + H$	R1
$+ H \rightarrow HS + H_2$	R2
$HS + HS \rightarrow H_2S + S$	R3
$\rightarrow H_2 + S_2$	R4
$HS + H + M \rightarrow H_2S + M$	R5
$HS + H \rightarrow H_2 + S$	R6
S and $S_2 \rightarrow S_8$ (elemental sulfur, yellow)	R7
$\rightarrow (NH_4)_x S_y$ (ammonium polysulfides, orange)	R8
$\rightarrow H_x S_y$ (hydrogen polysulfides, brown)	R9

The rate constant of reaction R2 is (Kurylo et al. 1971) $k_2 = (1.29 \pm 0.15)$ $\times 10^{-11} \exp[-(1709 \pm 60)/1.987\,T]$ cm^3 s^{-1}/molecule. Reaction R3 is more likely than R4.

[a] After Lewis and Prinn (1970).

explained by the various forms of sulfur and sulfur-bearing molecules produced in the H_2S photochemistry that could occur on Jupiter. Basically, in the relatively clear regions (such as the North Equatorial Belt of Jupiter) long wavelength UV photons could penetrate to the levels where H_2S might be present in the vapor phase. Photolysis of H_2S and subsequent reactions would eventually produce elemental sulfur (S_8), ammonium polysulfide $[(NH_4)_x S_y]$, and hydrogen polysulfide ($H_x S_y$), which are, respectively, yellow, orange, and brown in color. The complete chemical scheme according to Lewis and Prinn is given in Table 5.5.

There are several drawbacks to the above scenario of the H_2S photolysis at Jupiter. First, H_2S must be present in the upper troposphere/stratosphere region for it to be photolyzed. This is because H_2S photolysis occurs below 3170 Å, primarily around 2000 Å, as is evident from its photoabsorption characteristics in the UV (Fig. 5.8). As discussed earlier, solar photons shortward of 3000 Å are prevented from penetrating to levels below 1 bar due to Rayleigh scattering. The observations of Larson et al. (1984) place a severe upper limit of 3.3×10^{-8} on the H_2S mixing ratio in the troposphere (at 700 mb) of Jupiter. Most probably, H_2S is locked up in the NH_4SH cloud which is expected to form at 1.5 to 2 bar level on Jupiter (see Chap. 3 on Cloud Structure). Otherwise, H_2S should have been present in the solar composition ratio, 3.7×10^{-5}, in the troposphere, since its saturation vapor mixing ratio is greater even at the Jovian tropopause. In the unlikely event that H_2S were rapidly transported to the troposphere before its condensation, it may still not be feasible to form sulfur from its photolysis. A possibility exists that after the HS molecules – formed on H_2S photodecomposition – are transported down, they would undergo self-reaction, yielding H_2S_2 (Fowles et al. 1967), i.e.,

$$HS + HS \rightarrow H_2S_2^* \rightarrow H_2S_2 .$$

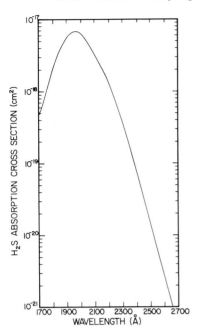

Fig. **5.8.** Photoabsorption cross-section of H$_2$S

The chemistry could then terminate without the formation of sulfur and the polysulfides, since H$_2$S$_2$ is highly stable in the Jovian atmosphere. The above arguments are consistent with the failure to detect H$_2$S in the atmosphere of Jupiter and seem to rule out sulfur in any form as the likely candidate for the chromophore of the clouds. Photolysis of H$_2$S at Saturn is even less likely. An attempt to detect H$_2$S in-situ in the deep troposphere of Jupiter will be made with the Galileo Probe Mass Spectrometer.

5.6 C – N Coupling: Methylamine (CH$_3$NH$_2$)

In a small altitude regime at around the 60 km altitude on Jupiter, the abundances of CH$_3$ and NH$_2$ are nearly equal. A possible reaction between these radicals would produce methylamine according to a chemical scheme given in Table 5.6. The rate of reaction R1 in Table 5.6, which produces CH$_3$NH$_2$, is comparable to the others in this table at around the 60-km altitude (the rate for reaction R1 is 1.1×10^{-10} cm^3 s^{-1} (Kuhn et al. 1977); rates for the other reactions appear in Tables 5.1 and 5.9). Model calculations indicate maximum CH$_3$NH$_2$ production rate of around 4×10^4 cm^{-3} (Jovian day)$^{-1}$, with a maximum volume mixing ratio of CH$_3$NH$_2$ of 2×10^{-11} (the earlier results of Kuhn et al. 1977 have been revised according to the new absorption cross-sections of CH$_4$ measured by Mount et al. 1977). The photochemical profiles of CH$_3$NH$_2$, CH$_3$, NH$_2$, and H, according to Kuhn et al. (1977), are shown in

Table 5.6. Photochemistry of CH_3NH_2[a]

Chemical reactions	Reaction number
$CH_3 + NH_2 \rightarrow CH_3NH_2$	R1
$CH_3 + CH_3 + M \rightarrow C_2H_6 + M$	R2
$CH_3 + H + M \rightarrow CH_4 + M$	R3
$NH_2 + NH_2 + M \rightarrow N_2H_4 + M$	R4
$NH_2 + H + M \rightarrow NH_3 + M$	

[a] After Kuhn et al. (1977).

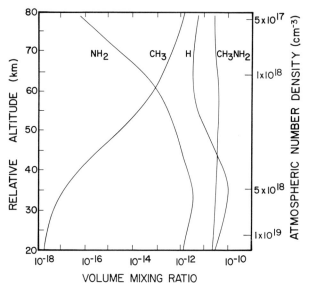

Fig. 5.9. Photochemical profiles of methylamine and the constituents important in its production. (Kuhn et al. 1977)

Fig. 5.9. The reaction of translationally hot hydrogen atoms with CH_4 to produce CH_3 (and subsequently CH_3NH_2, as in Table 5.6) has been included by Kaye and Strobel (1983a). The hot H atoms are formed on the photolysis of NH_3, PH_3, and C_2H_2. The resulting CH_3NH_2 mixing ratios are greater by about a factor of 3 compared to Kuhn et al. (1977). The CH_3NH_2 abundance is too low for it to be detected by the present instruments.

5.7 C−N Coupling: Hydrogen Cyanide (HCN)

A column abundance of 1.3×10^{17} cm^{-2}, with corresponding mixing ratio of 2×10^{-9} has been measured for HCN at Jupiter (Tokunaga et al. 1981). The

Table 5.7. HCN photochemistry[a]

Chemical reactions	Reaction number
$NH_3 + h\nu \rightarrow NH_2 + H$	R1
$H + C_2H_2 + M \rightarrow C_2H_3 + M$	R2
$NH_2 + C_2H_3 \rightarrow C_2H_5N$	R3
$C_2H_5N + h\nu \rightarrow HCN + CH_3 + H$	R4

[a] After Kaye and Strobel (1983b).

upper limit at Saturn is 6.5×10^{17} cm^{-2} (Tokunaga et al. 1981). The detection of this constituent in the infrared absorption implies that it should be present near and below the Jovian tropopause. Although Barshay and Lewis (1978) have shown that thermochemical equilibrium cannot account for its presence in the visible atmosphere, many plausible scenarios to explain its existence can be envisioned. First, photodissociation of CH_3NH_2 can directly lead to the formation of HCN, i.e.,

$$CH_3NH_2 + h\nu(\lambda > 1600 \text{ Å}) \rightarrow HCN + H_2 + 2H .$$

The above reaction is highly wavelength-dependent. Kaye and Strobel (1983b) calculate that an HCN abundance of 1×10^{17} cm^{-2} at the tropopause is possible, but a column abundance of 4×10^{16} cm^{-2} is more realistic. This is because of the low quantum yield (0.08, Gardner 1981) of HCN in the CH_3NH_2 photolysis. Moreover, the CH_3NH_2 abundance itself is quite low to begin with. The high temperature synthesis of atmospheric gases following lightning discharges would yield a mixing ratio of less than 3×10^{-12} for HCN, even if all of the internal heat were converted to lightning energy (Lewis 1980).

A feasible mechanism proposed by Kaye and Strobel (1983b) is the UV photodissociation of one of the isomers of C_2H_5N (cyclic aziridine) which is formed on reaction between NH_2 and C_2H_3 (Table 5.7). The kinetics and spectroscopy of only one of the four isomers of C_2H_5N, ethyleneimine or cyclic aziridine, have been studied in some detail. With available laboratory data, it is possible to arrive at a column abundance of 2×10^{17} cm^{-2} for HCN at the tropopause, which is comparable to the measured value. Since NH_3 photolysis at Saturn is severely retarded, only small amounts of NH_2, hence HCN would be produced by the above mechanism. Polymers of HCN have also been proposed as possible chromophores for the GRS and other clouds of Jupiter (Woeller and Ponnamperuma 1969).

5.8 Carbon Monoxide (CO)

Oxygen, the third most abundant element in the solar system, with O/H = 7×10^{-4}, is locked up predominantly in the water clouds at the major planets (see Chap. 2 on Cloud Structure). The visible atmosphere, therefore, is devoid of any large quantities of oxygen-bearing molecules. There has been one report of a possible H_2O detection in the stratosphere of Saturn using the International Ultraviolet Explorer (IUE) satellite (Winkelstein et al. 1983). However, highly extensive data set from Voyager UVS fail to substantiate the above finding. The only oxygen-bearing molecule detected so far is CO, definitely on Jupiter (Beer 1975; Beer and Taylor 1978; Larson et al. 1978), and tentatively on Saturn (Noll 1985). Beer and Taylor find a CO column abundance of $4.3 (+4.3, -2.2) \times 10^{17}$ cm^{-2}, volume mixing ratio 10^{-9}, and a rotational temperature of 125 ± 25 K. The measurements of Larson et al. (1978), on the other hand, yield a column abundance of 8×10^{17} cm^{-2} and a rotational temperature of $150-300$ K. CO is thermochemically stable at 1100 K, where it is produced on oxidation of CH_4 by H_2O. Prinn and Barshay (1977) have argued for a strong vertical mixing with $K = 2 \times 10^8$ cm^2 s^{-1} in Jupiter's interior to transport CO to the upper atmosphere. The wide discrepancy in the rotational temperatures of the above-mentioned observations of CO allows for the source to be either in the interior of the planet, as proposed by Prinn and Barshay, or extraplanetary, as discussed below.

An extraplanetary source proposed by Strobel and Yung (1979) involves removal of oxygen atoms from the Galilean satellites on impact of energetic charged particles. They calculated that Io would supply the largest flux of oxygen atoms into the atmosphere of Jupiter, approximately 10^6 to 10^7 cm^{-2} s^{-1}. This estimate was arrived at by assuming cosmic abundance ratio of Na to O. Na emissions were detected from a ground-based spectrometer (Brown 1974). Based on the laboratory experiments on the erosion of water-ice by MeV particles (Brown et al. 1978) and the available information on the Jovian magnetosphere, Lanzerotti et al. (1978) concluded that water-ice erosion at Galilean satellites should be appreciable. They estimate erosion rates of 3×10^8 to 1×10^{11} cm^{-2} s^{-1} at Europa, an order of magnitude smaller at Ganymede and approximately a factor of $(1.5-2) \times 10^3$ smaller at Callisto. If a mechanism can be found to remove this water to the atmosphere of Jupiter, the fluxes there would be scaled down by several orders of magnitude. The Voyager observations reveal, however, that the mechanism for removing oxygen from Io is quite different from that assumed prior to the Voyager observations. The total flux of oxygen from all Galilean satellites into Jupiter is derived to be on the order of $(0.3-5) \times 10^7$ cm^{-2} s^{-1} (Science 1979). The source of oxygen from Io lies in the photodissociation and charged particle dissociation of SO_2 (Kumar and Hunten 1982), which is released from the volcanoes of Io (Hanel et al. 1979a, b). Subsequent ionization of oxygen leads to oxygen ions, as is evident from their large concentrations in the Io torus (Broadfoot et al. 1979; Bridge et al. 1979). Some of the energetic oxygen atoms and ions are expected to enter Jupiter (see Chap. 6 on the Ionosphere for further discussion). Subsequently, they would lose their energy in collisions with H_2. The reactions leading up to the formation of CO are given in Table 5.8. The thermal oxygen atoms react with the methyl radicals formed in the CH_4 photochemis-

Table 5.8. CO photochemistry

Chemical reactions	Reaction number
$O + CH_3 \rightarrow HCHO + H$	R1
$HCHO + O \rightarrow HCO + OH$	R2
$HCO + O \rightarrow OH + CO$	R3
$\rightarrow H + CO_2$	R4
$HCO + H \rightarrow H_2 + CO$	R5
$O + CH_2 \rightarrow H_2 + CO$	R6

try, and produce formaldehyde, HCHO. After a series of further chemical reactions, CO and CO_2 are produced. It is not apparent in what form oxygen would enter the atmosphere. If it enters as oxygen ions (O^+), they too would eventually produce CO and CO_2, first by forming H_2O in several charge exchange and electron recombination reactions (see Chap. 6 on the Ionosphere for additional details). CO_2 is not stable in the upper atmosphere of Jupiter as it would be rapidly dissociated by the solar photons longward of 1600 Å. Oxygen ions can enter Jupiter along the magnetic flux tubes connecting Io to Jupiter, as was suggested for the transport of the sodium ions from Io to Jupiter by Atreya et al. (1974). Radial diffusion of oxygen is another possibility.

Another extraplanetary source of oxygen involving ablation of carbonaceous chondritic meteorites in the Jovian atmosphere has been proposed by Prather et al. (1978). The estimate for the flux of H_2O released is on the order of 10^7 to $2 \times 10^9 \, cm^{-2} \, s^{-1}$. Dissociation of water can supply oxygen atoms or hydroxyl radicals which undergo chemical reactions given in Table 5.8, and eventually result in the formation of CO, (OH reacts with CH_2, CH_3, and C_2H_2 to produce CO).

All of the above-mentioned sources are capable of providing the needed flux of around 4×10^7 O-atoms $cm^{-2} \, s^{-1}$ to explain the CO mixing ratio of 10^{-9} observed at Jupiter. To discriminate between them would require correct identification of the temperature-pressure regime of the origin of CO, and detection of oxygen and water in the upper atmosphere along with their relationship to the extraplanetary sources of O or H_2O. The extraplanetary source involving meteorites would predict presence of Si and SiO, whereas the Io source would predict diffusion into the atmosphere of S and the formation of CS. Finally, CO detection at Saturn in the 5 μm region (Noll 1985) places additional constraints on the source of CO on the major planets. (Note, however, that CO lines are strongly mixed in with PH_3 lines, whose spectroscopic data were not included in deriving the CO mole fraction at Saturn.) It appears that the lines are formed at around the 2-bar level. Both extraplanetary and the internal sources for CO are viable possibilities. An unreasonably high eddy diffusion coefficient of 10^9 to $10^{10} \, cm^2 \, s^{-1}$ in the troposphere may be needed to sustain the ppb levels of CO at Saturn, if strong vertical transport mechanism is invoked to explain CO. An attractive extraplanetary source at Saturn is water from the rings of Saturn, which is also proposed as a candidate needed for the ionospheric sink (see Chap. 6 on the Ionosphere). So far, water vapor has alluded detection in the visible atmosphere of Saturn or Jupiter. Charged particle sputtering of water-ice on Saturn's E-ring, and the moons Dione and Tethys can provide from 3×10^{22} to $4 \times 10^{24} \, s^{-1}$ oxygen ions following dissociation/ionization of the water (Cheng et al. 1982). Perhaps a more effective mechanism for the removal of water-ice is the impact of micrometeorites on the rings (Morfill et al. 1983). A fraction of the oxygen atoms, or water itself, so produced is expected to end up in Saturn's upper atmosphere and undergo chemical reactions, such as those in Table 5.8, to produce CO.

5.9 Methane (CH$_4$)

The photochemistry of methane in the atmospheres of Jupiter, Saturn, Uranus, and Neptune has been studied extensively by Strobel (1969, 1973 b, 1975); Atreya et al. (1981); Atreya (1982); Gladstone (1982); Atreya and Ponthieu (1983); Atreya (1984a) and Atreya and Romani (1985). CH_4 is photolyzed below 1450 Å (see Fig. 2.16 for the hydrocarbon absorption cross sections). For all practical purposes, however, CH_4 photolysis is done by the Lyman-α flux. The stable hydrocarbon products formed in the CH_4 chemistry are C_2H_2, C_2H_6, and C_2H_4. These hydrocarbons themselves undergo photolysis, mostly above 1450 Å, and can result in the formation of higher-order hydrocarbons such as butane, propane, polyacetylenes, benzene, etc. The CH_4 photochemistry relevant to the outer planets is schematically shown in Fig. 5.10. The important chemical pathways along with the associated rate constants are presented in Table 5.9.

Photolysis of CH_4 results in the formation of $^1CH_2$, $^3CH_2$, and CH. Direct production of CH_3 is, however, kinetically forbidden. The $^1CH_2$ radicals preferentially form CH_3 in the presence of H_2 as the background gas, as in Table 5.9, R3. If the background gas were not H_2 (e.g., N_2 on Titan), they would then be quenched to $^3CH_2$. The most probable reaction of $^3CH_2$ is with CH_3 to form C_2H_4 (R4). The other possibilities are: reaction with C_2H_2 forming C_3H_4 (allene $CH_2 = C = CH_2$, or methylacetylene CH_3C_2H), or a self-reaction forming C_2H_2. The most probable reaction of CH is with CH_4 forming C_2H_4 (R7); however, at low pressures it could react with H_2 to yield CH_3. The latter is not of great significance in the Jovian planets. As is evident from the chain of reactions R1 through R8, CH_4 photolysis rapidly produces C_2H_4. Photolysis of C_2H_4 produces C_2H_2 (R9 – R11), whereas its chemical reactions recycle some of the methane (R12 – R15). The self-reaction of the methyl radicals (CH_3) produced in the CH_4 photochemistry (R3), results in the formation of C_2H_6 (R16 – R19). C_2H_6 itself is photolyzed to yield C_2H_2 (R20 – R25) and C_2H_4 (R26). The latter undergoes chemical reactions to recycle CH_4, as discussed above. The acetylene molecules are extremely durable in the upper atmo-

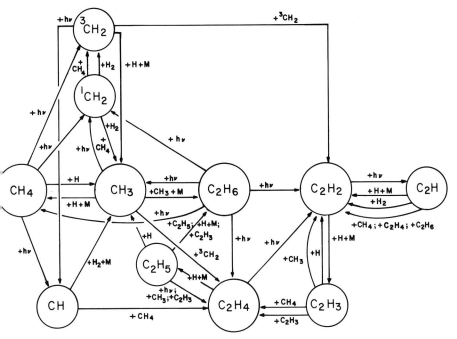

Fig. 5.10. Photochemical scheme for methane on the major planets

sphere, as is evident from the sequence of reactions R28 to R36. The C_2H_2 reactions R28 – R30 effectively recombine the H atoms catalytically. The C_2H_2 photolysis reactions (R31 – R33), on the other hand, in effect result in the H_2 dissociation in the upper regions where C_2H_2 is produced. In the lower atmosphere, the C_2H_2 photolysis in essence results in the catalytic dissociation of CH_4 (R34 – R36). In addition to the possible formation of heavier hydrocarbons (C_3H_8, C_4H_{10}, etc., R37 and R38), there is some likelihood of benzene (C_6H_6) formation by the following reaction (Okabe 1983).

$$C_2H_2^{**} + 2\,C_2H_2 \rightarrow C_6H_6 .$$

The triplet metastable state $C_2H_2^{**}$ is, however, expected to be rapidly deactivated by H_2 on Jupiter. Thus, the yield of C_6H_6 from the above reactions would be negligible. Allen and Yung (1985) have proposed the following alternate mechanism for the production of C_6H_6

$$C_2H_2 + C_2H_3 + M \rightarrow C_4H_5 + M$$

followed by

$$C_2H_2 + C_4H_5 \rightarrow C_6H_6 + H .$$

Detection of benzene with a mixing ratio of $2^{+2}_{-1} \times 10^{-9}$ in the north polar region (latitude $> 48°$, System III longitude $120° - 140°$) has been reported by

Table 5.9. CH_4 photochemistry[a]

Reaction	Quantum yield, q, or rate constant, k[b]		Reaction number	Reference
	q(at Ly-α)	q(other λ)		
1. CH_4 is photolyzed in the upper layers (max. J at Ly-α) and C_2H_4 is eventually produced (10% of all photons absorbed by CH_4 produce C_2H_4)				
$CH_4 + h\nu \rightarrow {}^3CH_2 + 2\,H$	0.51	0	R1	1, 2, 3, 4, 5, 6, 7
$CH_4 + h\nu \rightarrow {}^1CH_2 + H_2$	0.41	1.0	R2	1, 2, 3, 4, 5, 6, 7
${}^1CH_2 + H_2 \rightarrow CH_3 + H$	$k = 7.0 \times 10^{-12}$		R3	8
$CH_3 + {}^3CH_2 \rightarrow C_2H_4 + H$	$k = 7.0 \times 10^{-11}$		R4	8
Net $2\,(CH_4 + h\nu) \rightarrow C_2H_4 + 4\,H$			R5	
And				
$CH_4 + h\nu \rightarrow CH + H + H_2$	0.08	0	R6	1, 2, 3, 4, 5, 6, 7
$CH + CH_4 \rightarrow C_2H_4 + H$	$k = 1.0 \times 10^{-10}$		R7	9, 10
Net $2\,CH_4 + h\nu \rightarrow C_2H_4 + 2\,H + H_2$			R8	
2. C_2H_4 is quickly lost by photolysis to C_2H_2, or by recycling to CH_4				
$C_2H_4 + h\nu \rightarrow C_2H_2 + H_2$	0.51	0.51	R9	11, 12
$C_2H_4 + h\nu \rightarrow C_2H_2 + 2\,H$	0.49	0.49	R10	11, 12
Net $2\,(C_2H_4 + h\nu) \rightarrow 2\,C_2H_2 + H_2 + 2\,H$			R11	
And				
$C_2H_4 + H + M \rightarrow C_2H_5 + M$	$k_0 = 1.1 \times 10^{-23}\,T^{-2}e^{-1040/T}$ $k_\infty = 3.7 \times 10^{-11}\,e^{-1040/T}$		R12	13, 14
$C_2H_5 + H \rightarrow 2\,CH_3$	$k = 1.9 \times 10^{-10}\,e^{-440/T}$		R13	15, 16
$2\,(CH_3 + H + M \rightarrow CH_4 + M)$	$k_0 = 3.1 \times 10^{-29}\,e^{+457/T}$ $k_\infty = 1.5 \times 10^{-10}$		R14	17, 18, 10
Net $C_2H_4 + 4\,H \rightarrow 2\,CH_4$			R15	

Table 5.9 (continued)

3. Dissociation of CH$_4$ also leads to the formation of C$_2$H$_6$ (20% of all photons absorbed by CH$_4$ produce C$_2$H$_6$)

Reaction					
$2(CH_4 + h\nu)$ → $^1CH_2 + H_2$	0.41			R16	1, 2, 3, 4, 5, 6, 7
$2(^1CH_2 + H_2)$ → $CH_3 + H$				R17	8
$2CH_3 + M$ → $C_2H_6 + M$		1.00	$k = 7.0 \times 10^{-12}$ $k_0 = 1.0 \times 10^{-26}\,e^{+506/T}$ $k_\infty = 5.5 \times 10^{-11}$	R18	10, 19, 20
Net $2(CH_4 + h\nu)$ → $C_2H_6 + 2H$				R19	

4. C$_2$H$_6$ is also lost by conversion to C$_2$H$_2$ or by recycling to CH$_4$, as is the case with C$_2$H$_4$

Reaction				
$C_2H_6 + h\nu$ → $C_2H_2 + 2H_2$	0.27	0.25	R20	3, 21, 22, 23, 24
$C_2H_6 + h\nu$ → $C_2H_4 + H_2$	0.56	0.13	R21	3, 21, 22, 23, 24
$C_2H_6 + h\nu$ → $C_2H_4 + 2H$	0.14	0.30	R22	3, 21, 22, 23, 24

Followed by

$C_2H_4 + h\nu$ → $C_2H_2 + H_2$	0.51	0.51	R23	11, 12
$C_2H_4 + h\nu$ → $C_2H_2 + 2H$	0.49	0.49	R24	11, 12

Net $3C_2H_6 + h\nu$ → $3(C_2H_2) + 4H_2 + 4H$	R25

And

$C_2H_6 + h\nu$ → $C_2H_4 + H_2$	0.56	0.13	R26	3, 21, 22, 23, 24
→ $C_2H_4 + 2H$	0.14	0.30		

Followed by

C_2H_4 → CH$_4$ Conversion, (R15)	R27

$J_{C_2H_6} \leq 10\%\ J_{C_2H_4}$ so that C$_2$H$_6$ is stable in the upper atmosphere and is lost by diffusion to the lower altitudes.

5. Once produced in the above manner, C$_2$H$_2$ is highly stable. It undergoes the following reactions

Reaction			
$H + C_2H_2 + M$ → $C_2H_3 + M$	$k_0 = 6.4 \times 10^{-25}\,T^{-2}\,e^{-1200/T}$ $k_\infty = 9.2 \times 10^{-12}\,e^{-1200/T}$	R28	25
$C_2H_3 + H$ → $C_2H_2 + H_2$	$k = 1.5 \times 10^{-11}$	R29	26
Net $2H$ → H_2		R30	

And

Table 5.9 (continued)

Reaction		Quantum yield, q, or rate constant, k^b		Reaction number	Reference
		q (at Ly-α)	q (other λ)		
6.	$C_2H_2 + h\nu$	$\to C_2H + H$	0.40 at $\lambda < 1500$ Å	0.18 at $\lambda > 1500$ Å	R31	27, 28, 29
	$C_2H + H_2$	$\to C_2H_2 + H$	$k = 1.2 \times 10^{-10} e^{-2000/T}$		R32	30, 31
Net	H_2	$\to 2H$			R33	

C_2H_2 is therefore long-lived and its concentration builds up until its production is balanced by diffusive flux to the lower boundary. In the lower atmosphere also C_2H_2 is photolyzed and immediately recycled in the upper layer.

$C_2H_2 + h\nu$	$\to C_2H + H$	0.40 at $\lambda < 1500$ Å	0.18 at $\lambda > 1500$ Å	R34	27, 28, 29
$C_2H + CH_4$	$\to C_2H_2 + CH_3$	$k = 2.5 \times 10^{-11} e^{-750/T}$		R35	30, 31
Net	$CH_4 + h\nu$	$\to CH_3 + H$		R36	

This is an indirect manner in which CH_4 can be 'dissociated' to give $CH_3 + H$. The reaction $CH_4 + h\nu \to CH_3 + H$ is not permitted (Slanger 1982). CH_3 so produced, however, is either recycled to methane ($CH_3 + H + M \to CH_4 + M$) 65% of the time, or converted to ethane ($CH_3 + CH_3 + M \to C_2H_6 + M$) 35% of the time. Finally, higher order hydrocarbons, such as propane (C_3H_8), methylacetylene (C_3H_4), butane (C_4H_{10}), etc. are formed by reactions of the following nature:

$C_2H_5 + CH_3 + M$	$\to C_3H_8 + M$	R37
$C_3H_7 + CH_3 + M$	$\to C_4H_{10} + M$	R38

Heavier and complex hydrocarbons may also be produced during the charged particle or photochemical ionization processes.

[a] Adapted from Gladstone (1982); Atreya and Donahue (1979); Strobel (1975).

[b] Rate constants, k, in cm^3 s^{-1} for 2-body, and cm^6 s^{-1} for 3-body reactions. 3-body rates are given by: $k = \min (k_0, k_\infty/M)$ where $k_0 k_\infty$ are, respectively, the low and the high pressure limits of k, M is atmospheric number density.

[1] Watanabe et al. (1953); [2] McNesby and Okabe (1964); [3] Calvert and Pitts (1966); [4] Gorden and Ausloos (1967); [5] Rebbert and Ausloos (1972); [6] Mount et al. (1977); [7] Slanger (1982); [8] Laufer (1981a); [9] Butler et al. (1981); [10] Gladstone (1982); [11] Zelikoff and Watanabe (1953); [12] Back and Griffiths (1967); [13] Michael et al. (1973); [14] Lee et al. (1978); [15] Halstead et al. (1970); [16] Teng and Jones (1972); [17] Troe (1977); [18] Patrick et al. (1980); [19] van den Bergh et al. (1976); [20] Callear and Metcalfe (1976); [21] Akimoto et al. (1965); [22] Hampson and McNesby (1965); [23] Lias et al. (1970); [24] Mount and Moos (1978); [25] Payne and Stief (1976); [26] Keil et al. (1976); [27] Nakayama and Watanabe (1964); [28] Okabe (1981); [29] Laufer (1982); [30] Laufer (1981b); [31] Brown and Laufer (1981).

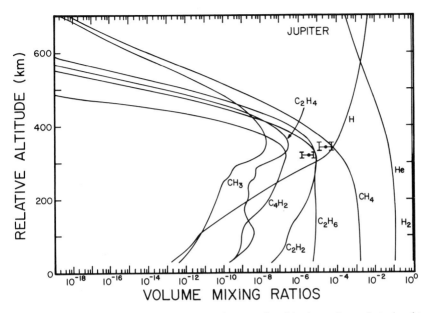

Fig. 5.11. Typical distribution of the hydrocarbons produced in the methane photochemistry on Jupiter. (After Gladstone 1982). The only data available for the photochemical regime are indicated

Kim et al. (1985). Detection of benzene and several other heavy hydrocarbons at the high latitudes would imply that the charged particle synthesis may be responsible for their production. Finally, polymerization of acetylene can result in the formation of diacetylene (C_4H_2) and polyacetylenes (additional details can be found in Chap. 7.4 on Titan Chemistry).

Polar brightenings of Jupiter seen in the infrared signatures of CH_4 (Caldwell et al. 1980, 1982), C_2H_2 (Drossart et al. 1986; Kim et al. 1985), C_2H_6 (Kim et al. 1985; Drossart et al. 1985; Kostiuk et al. 1985), C_3H_4 and C_6H_6 (Kim et al. 1985) can result from a combination of the enhanced stratospheric temperatures and enhanced abundances of the constituents. Either possibility can be explained by the charged particle precipitation at high latitudes.

The photochemical models constructed on the basis of the above chemistry (Atreya et al. 1981; Atreya 1982; Gladstone 1982; Waite et al. 1983) satisfactorily reproduce the available data on the hydrocarbons in the upper atmospheres of Jupiter and Saturn. A model for Jupiter's hydrocarbons, and the available data in the photochemical regime are shown in Fig. 5.11. At Jupiter, the height profiles of the hydrocarbons are available only near the homopause; the situation is even less satisfactory for the other major planets. To some extent, the data gap in the middle atmosphere is expected to be filled by the Galileo/Jupiter Entry Probe.

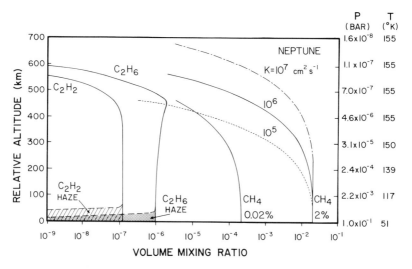

Fig. 5.12. Photochemical distributions of CH_4, C_2H_2, and C_2H_6 on Neptune. Altitudes are above the tropopause which is taken at $P = 100$ mb, $T = 55$ K level. CH_4 distributions are shown with $CH_4/H_2 = 2\%$ and $K = 10^5$, 10^6 and 10^7 cm^2 s^{-1}, and with $CH_4/H_2 = 0.02\%$ and $K = 10^6$ cm^2 s^{-1}. (Current data favor a 2 to 3% mixing ratio for methane.) Condensation of ethane and acetylene occurs within 50 km above the tropopause. The C_2H_2 and C_2H_6 distributions correspond to $CH_4/H_2 = 2\%$ and $K = 10^6$ cm^2 s^{-1}

The photochemistry of methane on Uranus and Neptune is directly coupled to the thermodynamics of C_2H_2 and C_2H_6. Due to the relatively low stratospheric temperatures on both these planets, condensation of C_2H_2 and C_2H_6 is highly likely over an extensive height range above the "cold trap" (tropopause), (Atreya and Ponthieu 1983; Atreya 1984a; Atreya and Romani 1985). In addition to the C_2H_2 and C_2H_6, formation and condensation of polyacetylenes is highly probable. The latter absorb sunlight out to longer ultraviolet wavelengths (up to 3200 Å) than either C_2H_2 or C_2H_6. The condensation scenario at Uranus is highly latitude-dependent due to its 98° equator to orbit inclination. At the Voyager/Uranus encounter epoch in 1986, e.g., the south pole pointed almost directly to the Sun (according to astronomical convention, the rotation pole south of the ecliptic plane is referred to as the south pole). For the Voyager epoch, therefore, larger solar radiation penetrates to deeper atmospheric levels at the pole than the equator, causing CH_4 dissociation to deeper levels in the south polar region. The resulting C_2H_2, C_2H_6, and the polyacetylenes would also condense at deeper levels than in the equatorial region. Scattering of sunlight off the deep, high density haze, or even the solid CH_4-ice cloud at the ~1.5-bar level can give Uranus a bright appearance. Indeed, ground-based observations indicate that the appearance of Uranus has steadily become brighter as the pole turned toward the Sun (Smith 1984; Smith B. A. 1983, personal communication). The presence of polar haze is clearly established from the violet filter (~4000 Å) images of Uranus taken by Voyager (Smith et al. 1986). In this wavelength region, the haze appears dark. (The south polar region appears dark also in the methane filter centered at around 6200 Å. This is due to the absorption by the deep and thick column of methane on Uranus. The absorption of red sunlight by methane is also responsible for Uranus' aquamarine (bluish-green) appearance in the clear filter). The methane filter images also give some indication of deep atmospheric haze (Owen T. C. 1986, personal communication). Once condensation of the hydrocarbon hazes has taken place, it is expected that some, if not all, of the haze particles will eventually snow out of the atmosphere.

A steady-state scenario for the C_2H_2 and C_2H_6 condensation, along with the hydrocarbon vapor distributions at Neptune is shown in Fig. 5.12. The stratospheric temperature gradient at Neptune is greater than at Uranus. It appears to be just enough to maintain sufficient abundance of C_2H_6 in the vapor phase for its detection at 12 μm, unlike at Uranus (Table 1.6). As on Uranus, there is some likelihood of polyacetylene haze in the stratosphere of Neptune. Because of Neptune's 27° inclination, the hydrocarbon condensation scenario is expected to be essentially opposite to that at Uranus, i.e., more hazy or cloudy conditions in the polar than in the equatorial region are expected. The discussion of hydrocarbon condensation is based on thermochemical equilibrium and steady state conditions. Effects of tropospheric/stratospheric dynamics, and latitudinal variations in temperature and the vertical mixing could alter the above-mentioned conclusions somewhat.

The present-day atmospheric mixing ratio of CH_4 on the major planets is at least as much as that expected for a solar composition atmosphere. On the other hand, photolysis destroys methane rapidly. The possible scenario for recycling methane is similar to that discussed earlier for ammonia. The heavier hydrocarbons diffuse down, some snow out of the atmosphere. Once in the high pressure, high temperature regime of the interior, they undergo pyrolysis, thus yielding CH_4 back. Methane is thus recycled, it then convects to the upper atmosphere.

5.10 Outstanding Issues

Photochemistry of nonequilibrium species such as PH_3, CO, GeH_4, SiH_4, AsH_3, HCN(?), etc. is not fully understood. Does strong vertical transport indeed exist in some regions of Jupiter and Saturn? Does the color of the various clouds of Jupiter and Saturn result from the photodecomposition of PH_3 or H_2S, or is it simply an optical effect? Why has H_2S not been detected? Is CO indigenous to the planet or does it have an extraplanetary origin? What is the abundance of N_2 in the atmospheres of Jupiter or Saturn; is it indicative of a photochemical or an internal source? Do charged particles (including lightning) modify the distribution and make-up of the hydrocarbons? Are they indeed the cause of polar infrared brigthenings? What is the distribution of photochemically active constituents in the middle atmosphere? What role do the hot hydrogen atoms play in photochemistry? How are photochemical processes affected by the presence of hazes and clouds? What are the photoabsorption properties of the various photochemically generated constituents in condensed phase? The methods outlined earlier for resolving the many outstanding issues in Chapters 1 and 2 on Composition and Thermal Structure are equally applicable here. In addition, models including the effects of charged particle precipitation and condensation will be needed.

6 Ionosphere

6.1 Measurements

The ionospheric structure of the major planets has been measured since 1973, using the technique of radio occultation (Chap. II) from Pioneer and Voyager spacecrafts. The principal difference between the Pioneer and the Voyager measurements is that the former were carried out using a single frequency oscillator (S-band at 2.293 GHz or 13 cm-λ), while the latter employed dual frequency mode (S-band, and X-band at 8.6 GHz or 3.6 cm-λ). The dual frequency mode is especially powerful for the interpretation of the Jupiter and

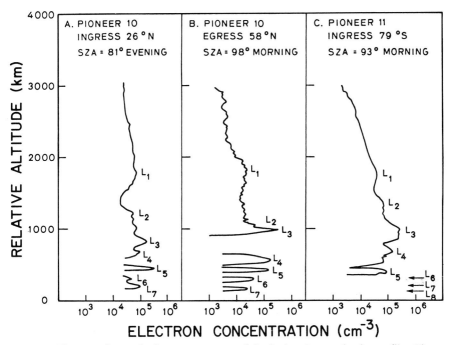

Fig. 6.1. Pioneer radio occultation measurements of the Jovian electron density profiles. Figure re-drawn by Chen (1981) on a common height scale with zero altitude at 10^{19} cm^{-3} atmospheric density level, which lies approximately 62 km above the 1-bar level and 50 km above the NH_3-cloud tops

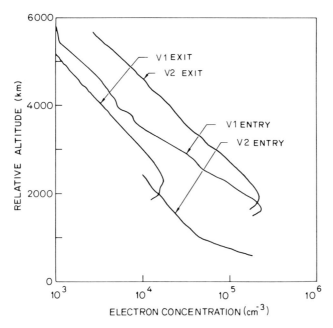

Fig. 6.2. Voyager radio occultation measurements of the Jovian electron density profiles. Zero altitude is at the 1-mb pressure level which is 160 km above the NH_3-cloud tops. (Courtesy of the Voyager Radio Science Team)

Saturn ionospheric data which are replete with effects of multipath propagation of the radio beam in sharp ionization layers. Moreover, greater sensitivity has been achieved in the Voyager data, so that the signal-to-noise ratio exceeds the Pioneer 13 cm-λ value by 10 dB and 23 dB at the S and X-bands, respectively (Eshleman et al. 1977). Reasonably good quality data were obtained on all except the Pioneer 11 egress (also called exit or emersion) occultation.

The Pioneer observations of Jupiter's ionosphere, presented in Fig. 6.1, show considerable latitudinal, diurnal, and temporal variations in the location and magnitude of the peaks and the plasma scale heights. Nevertheless, it seems that the maximum electron concentration in all these observations is between 1×10^5 and 2×10^5 cm^{-3}, and it occurs at an altitude of approximately 1000 km above the 1-bar level. Up to eight ionization layers (L1 − L8) are also recorded above and below the peak. Layers L6 and L7 may not be real ionization layers, as they were perhaps caused by deviations from spherical symmetry or scintillation noise (Fjeldbo et al. 1975, 1976). The topside plasma scale heights and the corresponding plasma temperatures appear to be highly variable. Under assumptions of thermal equilibrium between electron, ion, and neutral temperatures, and the major ion as H^+, the topside temperatures, however, hover around 800 K.

The Voyager measurements of the Jovian ionosphere were done close to the solar maximum, unlike the Pioneer data, which correspond to the solar minimum. The Voyager observations are presented in Fig. 6.2. The analysis of the dual frequency data for the lower ionosphere where presumably the ionospheric layers are present (see, e.g., Fig. 6.1 for Pioneer), is highly complex and as yet incomplete. The Voyager 1 data are for the near-equatorial latitudes, and seem to indicate a factor of 10 decrease in the "peak" electron concentration from ingress (also called entry or immersion) to egress. A word of caution is in order here and throughout the entire discussion of the Voyager ionospheric data. Until the lower ionospheric data have been analyzed, it is not certain where the "main peak" of the electron profile is located, since many ionization layers could equally give the appearance of main peak, as seen in the Pioneer data (Fig. 6.1). The above-mentioned decrease in the Voyager 1 electron concentration has been variously attributed to the diurnal change, or the longitudinal asymmetry. The Voyager 2 data are for relatively high latitudes. The entry occultation results give an indication of the ionization taking place to deep atmospheric levels. The details of all Jupiter ionosphere observations are summarized in Table 6.1. In this table, magnetic dip angle, I, is the inclination of the magnetic field relative to the surface (or cloud tops in the case of the outer planets), and λ_{III} is the System III (1965) longitude [1]. As in the Pioneer data, the Voyager observations also show large variations with latitude, longitude, etc. Furthermore, there is little consistency with the Pioneer data.

The Saturn ionospheric measurements have been carried out from Pioneer 11 (also called, Pioneer/Saturn) and the Voyager 1 and 2 spacecrafts. All observations were done near solar maximum conditions. The radio occultation occurred close to the terminator, i.e., with $\chi = 90°$. The Pioneer entry occultation data are more reliable than the exit data, as the latter suffered from the oscillator drift problems. The analysis of the Voyager 1 exit data is poorly understood, and is as yet incomplete. A composite of all Saturn ionospheric observations is shown in Fig. 6.3.

The Pioneer data are for the equatorial region. The peak electron concentration is on the order of $10^4 \, \text{cm}^{-3}$, occurring at an altitude of around 1800 km in the entry, and 1000 km higher in the exit data. Electron concentrations of

---

[1] The System III describes the rotation of Jupiter's magnetic field. The present convention is to use System III (1965) longitudes, λ_{III} (1965), which are related to the earlier 1957, January 1 longitudes, λ_{III} (1957.0), as follows

$$\lambda_{III} (1965) = \lambda_{III} (1957.0) - 3.04 \, T \, ,$$

where T represents time expressed in years (including fractions of a year) since 00 UT January, 1, 1957. The rotation period for 1957 was adopted as 9 h 55 m 29.37 s on the basis of then available radio emission data. The new data indicate that the 1957 period is too short by ~ 0.35 s, which would result in a 3°/yr longitude error of a magnetically related phenomena. System III (1965) is the latest attempt at defining the Jovian longitudes accurately. For further details see Dessler (1983).

Table 6.1. Jupiter ionosphere observations

	Pioneer				Voyager			
	10		11		1		2	
	Ingress	Egress	Ingress	Egress	Ingress	Egress	Ingress	Egress
Date	1973 December 4		1974 December 3		1979 March 5		1979 July 10	
Latitude, l	26°N	58°N	79°S	20°N	12°S	1°N	67°S	50°S
Longitude, λ_{III}	91°W	260°W	239°W	61°W	63°W	314°W	255°W	148°W
Solar zenith angle, χ	81°	95°	93°	79°	82°	98°	88°	92°
Time of day	Evening terminator	Morning terminator	Morning terminator	Afternoon	Afternoon	Predawn	Evening	Dawn
Dip angle, I	$-27°$	$-52°$	88°	$-8°$	45°	$-16°$	78°	76°
Plasma scale height, H_P [km]	675 ± 300	800	540 ± 60	Poor quality data	960 km above 3500 km; 590 km below 3500 km	960	1040	880
Ion temperature,[a] [K]	900 ± 400	1100	850 ± 100	Poor quality data	1150, 760	1190	1600	1200
Height of layer L1 [km]	1800 – 2000	1600 – 2000	1600	1800	1600	2300	≤600	1900
Height of layer L3 [km]	800	900	1100	800	?	?	?	?
Electron conc. at L1 [cm^{-3}]	$(3-9) \times 10^4$	2×10^4	4×10^4	?	$(2.2) \times 10^5$	1.8×10^4	$\geq 2 \times 10^5$	2×10^5
Electron conc. at L3 [cm^{-3}]	$(1-2) \times 10^5$	4×10^5	2×10^5	?	?	?	?	?

[a] Assuming $T_e = T_i = T_n$, and H^+ as the dominant ion.

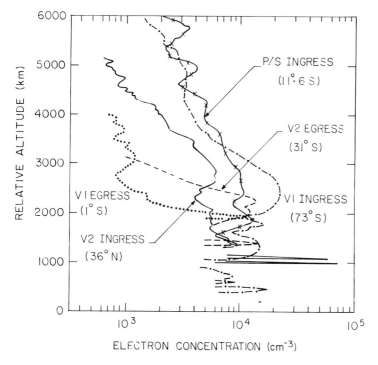

Fig. 6.3. Voyager (V1 and V2) and Pioneer (P/S) radio occultation measurements of the Saturn electron density profiles. Zero altitude is at the 1-bar pressure level. The sharp ionization layers in the lower ionosphere are similar to those seen in the ionosphere of Jupiter by Pioneer (Fig. 6.1). (After Lindal et al. 1985 and Kliore et al. 1980a)

less than 3×10^3 cm^{-3} may be due to electron fluctuations of the interplanetary solar wind, and not indicative of the planetary ionospheric concentrations (Kliore et al. 1980a, b). In any event, the ionosphere appears to be rather extensive. The lack of adequate information about the topside in these data renders the derivation of the plasma scale height virtually meaningless. Nevertheless, a range of 500 to 1000 K for the topside electron/ion temperatures appears likely.

The Voyager 1 and 2 measurements of the Saturn ionosphere, presented in Fig. 6.3, were done at both the low and high latitudes. The Voyager 1 occultation occurred between 73°S and 78.5°S; the ionospheric measurements, however, were done at the beginning, i.e., near 73°S. This is an important fact, as the auroral electron precipitation zone lies between 78° and 81°. The Voyager 2 data were taken in the equatorial to midlatitude region. The peak electron concentrations in the Voyager data are found to lie between 6×10^3 and 2×10^4 cm^{-3}. The topside plasma scale height (hence the electron temperature) derived from the Voyager 1 data is, surprisingly, lower than that from any other Saturn occultation measurements. As with Jupiter, the lower

Table 6.2. Saturn ionosphere observations

| | Pioneer/Saturn (S-band) | | Voyager (S and X bands) | | | |
| | | | 1 | | 2 | |
	Ingress	Egress	Ingress	Egress	Ingress	Egress
Date	1979 September 1		1980 November 12		1981 August 26	
Latitude, l	11.6°S	9.7°S	73°S	1°S	36°N	31°S
Solar zenith angle, χ	89.2°	90.9°	89°	~90°	87°	93°
Time of day	Terminator	Terminator	Late afternoon	Morning terminator	Late afternoon	Predawn
Plasma scale height, H_P [km]	See text	See text	560	~600	1000, topside 260, lower	1100, topside
Ion temperatures, [K]	See text	See text	~300	~400	565 K at 2800 km	617 K at 2500 km
Height of electron "peak" above 1 bar, [km]	1900	2900	2500	1900	2850	2150
Maximum electron conc. [cm^{-3}]	1.1×10^4	~1×10^4	2.3×10^4	1.1×10^4	6.4×10^3	1.7×10^4

ionospheric data collected on Voyager are still in the process of being analyzed. The principal characteristics of all the Saturn ionospheric observations are summarized in Table 6.2.

An indirect measure of the diurnal variation of maximum electron concentration at Saturn was inadvertently provided by the Planetary Radio Astronomy (PRA) experiment on Voyager. PRA detected sporadic bursts of $30 - 250$ ms duration in its entire frequency band of 20 kHz to 40 MHz (Evans et al. 1981; Warwick et al. 1981). These were termed Saturn Electrostatic Discharges (SED). The SED's have a periodicity of 10 h 10 $\pm$ 3 m, distinctly different from the planetary rotation rate of 10 h 39 m. The latter, however, refers to the planetary interior, whereas the rotation rate at the cloud tops is found to be 10 h 14 m from ground-based and Voyager imaging data. The rings at 1.8 R_s also rotate with a period of 10 h 10 m. Several characteristics of the SED's seem to favor their origin in the planetary atmosphere rather than in the rings – they are presumably caused by a planetary lightning storm (Kaiser et al. 1983; Burns et al. 1983). The frequency of the radio waves associated with the SED's must be greater than the electron plasma frequency, f_p,

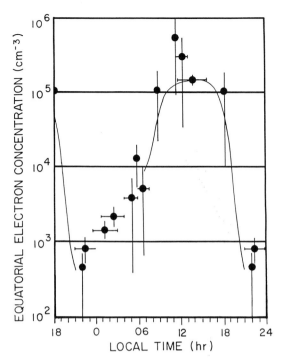

Fig. 6.4. Nighttime peak electron concentrations at Saturn derived from the Voyager PRA data. (After Kaiser et al. 1984 and Connerney and Waite 1984)

to permit their propagation through the ionosphere. f_p is given by the following expression (Spitzer 1962).

$$f_p = (N_e/\pi m_e)^{1/2}e \tag{6.1a}$$

or

$$f_p [\text{kHz}] = 9N_e^{1/2} , \tag{6.1b}$$

where N_e is the maximum electron concentration in cm^{-3}.

The above expression yields 0.7 MHz and 1.4 MHz for the cut-off frequencies corresponding to the observed range of 0.64×10^4 to $2.3 \times 10^4 \text{ cm}^{-3}$ in N_e. These electron concentrations are, however, for the dayside (terminator) conditions. Kaiser et al. (1984) have exploited the radio frequency information of the SED data to arrive at the maximum allowable ionospheric concentrations in the manner just outlined. The result is a diurnal variation in N_e as shown in Fig. 6.4. The maximum in N_e occurs around noon, the minimum occurs after dusk. This seems logical for an EUV-driven ionosphere which has short-lived ions. Although the error bars are large, there is an indication in Fig. 6.4 that at night there is a slow rise in N_e, which is puzzling. Following the terrestrial analogy, Connerney and Waite (1984) invoke a mechanism which would pump protons outwards during the day, and return them to the ionosphere at night.

The radio science experiment on Voyager 2 successfully probed the ionosphere of Uranus in January 1986 (Tyler et al. 1986). The entry and exit occultations occurred close to the equator (2° and 6° latitudes). Preliminary indications are that the ionosphere is quite extensive, with at least two widely spaced discrete ionization layers. Peak electron concentration of several thousand cm^{-3} is estimated. Measurements of the Neptune ionosphere are expected in August 1989, again using the technique of radio occultation on Voyager 2.

6.2 Theoretical Considerations

The radio occultation observations are incapable of identifying the ions; for that, in-situ measurements such as those with an ion mass spectrometer are necessary. It is, however, possible to construct reasonable scenarios for the ionospheric processes on the major planets by taking into account production and loss of ions and solving the appropriate conservation equations. Only a handful of constituents of the major planets can be directly ionized. These are H_2, He, and H. Possibility also exists for the ionization of some hydrocarbons, most notably the methyl radical, CH_3, and perhaps even CH_4. Metals such as Na, if they are present in the atmosphere, would also undergo ionization in a narrow atmospheric region. The ionization potentials (i.e., energy or wavelength at ionization threshold) of some constituents that are likely to be important in the ionospheres of the major planets are listed in Table 6.3.

6.2.1 Continuity Equation

The distribution of electrons and ions is governed by the continuity equation of the following form

$$\frac{\partial n_x}{\partial t} = P_x - L_x - \nabla \cdot \phi \ , \tag{6.2}$$

where the left-hand term represents the rate of change of the electron or ion concentration, P_x and L_x are, respectively, the ion or electron production and loss rates per unit volume, and ϕ is the diffusive flux. In the literature, loss term is sometimes written as $l_x n_x$, where l_x is expressed in units of s^{-1}. In addition to Eq. (6.2), the condition of charge neutrality of the plasma must be satisfied, i.e.,

$$n_e = \sum_i n_i \ , \tag{6.3}$$

where symbols e and i stand for ions and electrons, respectively. Also, for steady state, $\partial n_x / \partial t = 0$. Depending upon the lifetimes, either the chemical or the diffusive processes would control the loss, and hence the distribution, of the ions. For chemical equilibrium to prevail, the production and loss terms

Table 6.3. Ionization potentials

Constituent	E_{IP} [eV] [a]	λ_{IP} [Å] [a]
H	13.59	912
H_2	15.41	804
He	24.58	504
CH_3	9.82	1262(?)
CH_4	13.00	954
Na	5.14	2410
Si	8.15	1520
NH_3	10.15	1221
H_2O	12.60	985
C	11.3	1100
O	13.61	911
N	14.54	852
N_2	15.58	796
S	10.36	1196
Ar	15.75	787

[a] E_{IP} [eV] is the ionization potential in electron-volt, the corresponding wavelength in Angstrom is λ_{IP}. $1\,eV = 1.60206 \times 10^{-12}$ erg. The breaks in the table indicate decreasing order of relevance to the major planets.

must balance each other, i.e., $P = L$, and the diffusion should be negligible. The relevant chemical equilibrium time constant, also called chemical lifetime, τ_c, is given by

$$\tau_c = n_i/L_i .\tag{6.4a}$$

Chemical equilibrium is usually a good assumption up to the altitude of the main peak in the electron profile.

In diffusive equilibrium, the flux divergence term vanishes, and chemistry is no longer important. The diffusive equilibrium time constant, also called diffusive lifetime, τ_D, is given by

$$\tau_D = H_D^2/D_a ,\tag{6.4b}$$

where H_D is the scale height of the background gas through which the ion diffuses, and D_a is the ambipolar diffusion. D_a is further expressed as

$$D_a = kT_p/m_i \nu_{in} ,\tag{6.5}$$

where $T_p = T_i + T_e$, is the plasma temperature (i.e., the sum of electron and ion temperatures), m_i is the ion mass, and ν_{in} is the ion-neutral collision frequency. Ambipolar diffusion constrains the electrons and ions to diffuse with equal speed, and it results from an electric field. Such a polarization electric field is necessary, since in its absence charge separation between the faster-moving electrons and the slower-moving ions would take place. For equilib-

Table 6.4. Polarizability

Constituent	$\alpha\,[10^{-24}\,\mathrm{cm}^3]$
H_2	0.82
H	0.667
He	0.21
N_2	1.76

rium between the electron and ion temperatures, the plasma scale height is twice the ion scale height. The ambipolar diffusion would then be equal to twice the ion diffusion coefficient. The ion-neutral collision frequency term in Eq. (6.5) is given approximately by the following expression (for the ion and neutral temperatures below 300 K)

$$\nu_{in} = 2\pi(\alpha e^2/\mu_{in})^{1/2}n = 3.02 \times 10^{-9}(\alpha/\mu_{in})^{1/2}n \ , \tag{6.6}$$

where n is the concentration of the background through which the ion diffuses, $\mu_{in} = (1/m_i + 1/m_n)^{-1}$ is the reduced mass, and α is the polarizability factor. The value of α for various relevant constituents is listed in Table 6.4.

In the topside ionospheres of the major planets where diffusion of H^+ through H is expected, the following value of ν_{in} is obtained [using Eq. (6.6) and Table 6.4],

$$\nu_{H^+ - H} = 2.7 \times 10^{-9}n\,\mathrm{s}^{-1} \tag{6.7}$$

(for ion and neutral temperatures below 300 K).

The above value, however, needs to be adjusted when there is symmetric charge exchange taking place, e.g., between H^+ and H.

In that case,

$$\nu_{X^+ - X} \propto (T_i + T_n)^a \ , \tag{6.8}$$

where a varies between 0.3 and 0.4 (Stubbe 1968). At the Jovian exospheric temperature (~ 1000 K), the following value of the ion-diffusion coefficient for H^+ diffusing through neutral H is found (assuming $T_i = T_e = T_n$ so that $D_{H^+ - H} = 1/2\,D_a$), by using Eqs. (6.8) and (6.5).

$$D_{H^+ - H} \cong 1.9 \times 10^{19}/n\,\mathrm{cm}^2\,\mathrm{s}^{-1} \tag{6.9}$$

(rather than $3 \times 10^{19}/n$, the value without the above-mentioned correction in $\nu_{H^+ - H}$ due to the symmetric charge exchange).

With the knowledge of production, loss, and diffusion rates, the continuity equation for the ions [Eq. (6.2)] can be solved numerically to determine the ion profiles. The sources and sinks for the ions are discussed below.

6.2.2 Ionization Sources

In the equatorial region, ion production is dominated by the solar EUV photons despite their diminished flux at the outer planets (the only exception is perhaps a small localized longitude region on Jupiter, called the equatorial hydrogen Lyman-α bulge (Sandel et al. 1980), where particle precipitation may be important). At high latitudes, ionization due to energetic charged particle precipitation is expected to be dominant. The charged particles in this context comprise electrons, protons, or heavy ions such as those of sulfur and oxygen atoms. Indeed, analysis of the Voyager Cosmic Ray Subsystem data on the $1-20$ MeV/nuc oxygen, sulfur and sodium ions shows an inwardly directed density gradient between 6 and 17 R_J (Gehrels and Stone 1983). The power input implied from the precipitation of these ions into the Jovian atmosphere is of the order of 10^{12} to 10^{13} W. The amount of power is about what is required to explain the Jovian auroral emissions. The source of these energetic sulfur and oxygen atoms/ions is presumably related to the volcanic outgassing of SO_2 from Io (see Chap. 7 on Satellites). The sulfur and oxygen atoms are ionized by electron impact at a rate of $(2-20) \times 10^{28}$ s^{-1} in the Io plasma torus. The S and O ions are removed to the Jovian ionosphere either following acceleration in the Jovian magnetosphere, with subsequent sweeping by the rapidly rotating Jovian magnetic field, or they could diffuse radially by flux tube interchange (Siscoe and Chen 1977). In addition to the above two ionization mechanisms, cosmic rays with energies on the order of 10^9 eV (1 BeV) might play a role in the ionization at deep atmospheric levels, i.e., at pressures greater than 1 atm. Peak ionization rate of approximately 10 cm^{-3} s^{-1} has been estimated due to this process (Capone et al. 1979). Lower energy cosmic rays (10 to 100 MeV protons) may result in polar cap ionization, as in the terrestrial analog. The method of calculating the ion production rate from the two main processes (EUV and charged particles) is discussed below.

Photoionization

The production rate of an ion at altitude z following absorption of photons with wavelength $\lambda (\lambda < \lambda_{IP})$ is given by the following expression

$$p(z, \lambda) = \sigma_I(\lambda) F(z, \lambda) n(z) \qquad (6.10)$$
or
$$p(z, \lambda) = \sigma_I(\lambda) \{F_\infty(\lambda) \exp[-\tau(z, \lambda)]\} n(z) \ , \qquad (6.11)$$

where $\sigma_I(\lambda)$ is the ionization cross-section, and $n(z)$ is the number density of the constituent undergoing ionization (see Chap. 5 on Photochemistry for other details). The ion-production rate due to the ionizing flux at all wavelengths is

$$P(z) = \sum_\lambda p(z, \lambda) \ . \qquad (6.12)$$

Since ion production calculations are done numerically, the concepts can be best illustrated here by considering monochromatic radiation, i.e., by dropping λ in the above expression. It will then follow, that at height z

$$p = \sigma_I F_\infty \exp(-\tau) n \ . \tag{6.13}$$

Using definitions for τ and the slant path (from Chap. 5) and by setting $dp/dz = 0$, the following condition for the maximum ion production rate is obtained

$$\sigma_a H n(z_m) \sec \chi = 1 \ , \tag{6.14}$$

where H is the scale height, considering an isothermal atmosphere, (this is a reasonable assumption at the height of the electron peak), and z_m is the altitude of maximum ion production. Furthermore, for overhead sun, $\sec \chi = 1$, the maximum ion production rate would occur where

$$\sigma_a H n(z_m) = 1 \tag{6.15}$$

i.e., where

$$\tau = 1 \ . \tag{6.16}$$

The value of z_m can be obtained from Eq. (6.14) by taking exponential variation in density, i.e.,

$$n(z_m) = n_0 \exp(-z_m/H) \ . \tag{6.17}$$

Thus

$$z_m = H \ln[H \sigma_a n_0 \sec \chi] \ . \tag{6.18}$$

Similarly, the maximum ion production rate is obtained by substituting $n(z_m)$ from Eq. (6.14) in Eq. (6.13), i.e.,

$$p_m = (\sigma_I/\sigma_a) F_\infty \exp(-1)/H \sec \chi = 0.37 (\sigma_I/\sigma_a) F_\infty \cos \chi/H \ . \tag{6.19}$$

As discussed earlier, the maximum in ion-production rate occurs at an altitude where the optical depth is unity.

Charged Particle Ionization

Energetic charged particles at auroral latitudes can result in the production of large ion abundances. The energy of electrons and protons responsible for the auroral excitation at the outer planets lies generally between 1 and 100 keV. The rate of production of ions following the charged particle impact of the neutral gas is given by the following expression (Singer and Maeda 1961; Bauer 1973)

$$p = \frac{\rho(z)}{0.032} \int_{E_1}^{E_2} L(E) j(E, z) dE \ , \tag{6.20}$$

where $\rho(z)$ is the atmospheric mass density, $L(E)$ is the energy loss term (estimated empirically), $j(E, z)$ represents the energy spectrum of the ionizing particles, and E_1 and E_2 represent the energy range of the incoming particle beam.

For extremely large energy particles, such as cosmic rays, the ion production rate is given by an expression of the following type

$$p = \frac{1}{E_p} \int_E \int_z dE/dx \, j(E) \, dE \, d\Omega , \qquad (6.21)$$

where E_p is the average energy for an ion-pair production (approximately 35 eV for incident energies ≥ 50 eV), Ω is the solid angle, and dE/dx is the energy loss per unit distance. In both Eqs. (6.20) and (6.21), the differential energy spectrum on top of the atmosphere, $j(E)$, must be estimated, as its measurements at the major planets are not available.

6.2.3 Ion Loss

Once produced, the ions undergo chemical and diffusive losses. In the ionospheres of the major planets, the important chemical sinks are due to charge exchange with atmospheric constituents, and recombination. These processes and the associated time constants are discussed below.

i. Charge Exchange

This sink is also referred to as charge transfer, since typically it involves transfer of charge from an ion A^+ to a neutral B, i.e.,

$$A^+ + B \xrightarrow{k} B^+ + A , \qquad (6.22)$$

where k is the rate coefficient (in $cm^3 s^{-1}$) of the above reaction.
The rate of loss of ion A^+ is given by

$$\frac{\partial A^+}{\partial t} = -k[A^+][B] , \qquad (6.23)$$

where the square brackets represent number densities (in cm^{-3}). Integration of the above expression yields the time required, T_{ch}, for the change in ion concentration from an initial value $[A_0^+]$ to some value $[A_1^+]$, so that

$$T_{ch} = \frac{1}{k[B]} \ln \left(\frac{[A_0^+]}{[A_1^+]} \right) . \qquad (6.24)$$

The charge exchange time constant [or lifetime of reaction in (6.22)], τ_{ch}, is then the time during which the ion concentration drops by a factor of e, i.e.,

$$[A_1^+]/[A_0^+] = e^{-1} . \qquad (6.25)$$

Hence

$$\tau_{ch} = \frac{1}{k[B]} . \qquad (6.26)$$

For 3-body reactions, such as those in the denser part of the atmosphere,

$$\tau_{ch} = \frac{1}{k[B][C]} \; , \tag{6.27}$$

where k is expressed in units of $cm^6 s^{-1}$. Charge transfer loss mechanism could also involve molecules, so that there is an atom-ion interchange, e.g.,

$$A^+ + BC \xrightarrow{k} AB^+ + C \tag{6.28}$$

or it could involve dissociative charge transfer, e.g.,

$$A^+ + BC \rightarrow B^+ + A + C \; . \tag{6.29}$$

In both cases, the τ_{ch}'s are calculated in the same manner as for Eq. (6.22).

ii. Recombination

The most efficient mechanism for the loss of terminal ions in most planetary ionospheres is recombination with electrons. When neutralization of atomic ions takes place due to this process, it is termed radiative recombination. Neutralization of molecular ions is called dissociative recombination. The time constants for recombination can be derived from consideration of the following reaction

$$A^+ + e^- \xrightarrow{\alpha_r} A^* + h\nu \; , \tag{6.30}$$

where α_r is the radiative recombination coefficient (in $cm^3 s^{-1}$). The radiative recombination lifetime τ_r is then obtained in the same manner as τ_{ch} [Eq. (6.26)], i.e.,

$$\tau_r = \frac{1}{\alpha_r[N_e]} \; , \tag{6.31}$$

where $[N_e]$ is the electron concentration (in cm^{-3}). For most atomic ions, α_r is on the order of a few times $10^{-12} cm^3 s^{-1}$, whereas the peak electron concentrations in planetary atmospheres are around $10^4 - 10^5 cm^{-3}$. Hence, the lifetime of the ion against radiative recombination loss would be in the neighborhood of 10^7 to 10^8 s. In the topside ionospheres, therefore, atomic ions are extremely durable. Dissociative recombination loss, e.g., for reactions of the following type

$$AB^+ + e^- \xrightarrow{\alpha_d} A + B \tag{6.32}$$

is, on the other hand, extremely rapid, since the associated rate constants are on the order of 10^{-6} to $10^{-8} cm^3 s^{-1}$. A molecular ion, therefore has a relatively short lifetime.

When both positive and negative ions are present, recombination between them results in mutual neutralization. Typical rate constants for such reactions are $\sim 10^{-7} cm^3 s^{-1}$. Negative ions are generally formed in denser parts of planetary atmospheres (e.g., D region on Earth) by attachment of electrons to neutral constituents; the associated typical rate constants are $\sim 10^{-15}$ cm^3 s^{-1}. No measurements of the part of Jovian ionospheres where such ions may be present are yet available.

6.2.4 Ion Chemistry

A crude estimation of the electron profile in the outer planet ionospheres may be obtained by considering a simplified chemical scheme, presented earlier in the context of hydrogen chemistry (Fig. 5.1). Even the chain of reactions beginning with He and ending in H_3^+ (Fig. 5.1) can be ignored, since it contributes less than 20% to the electron concentration. In the absence of charged particle impact, absorption of solar EUV radiation shortward of 804, 912, and 504 Å, respectively, by H_2, H, and He leads to the production of primary ions. The relevant cross-sections are shown in Fig. 6.5. The solar EUV fluxes are known to vary by up to factor of 3 between solar maximum and solar minimum. The solar fluxes near the last solar maximum when the Voyager observations of Jupiter and Saturn occurred are shown in Fig. 6.6a. The ratios of these fluxes to those near the solar minimum are presented in Fig. 6.6b. While there is general agreement between the above-mentioned flux

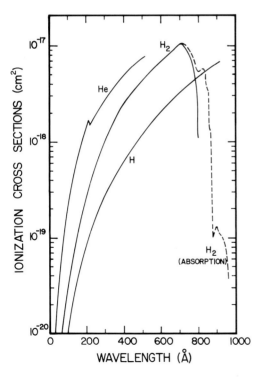

Fig. 6.5. Ionization cross-sections of H_2, H and He. For H_2, absorption cross-section beyond the ionization threshold is also shown

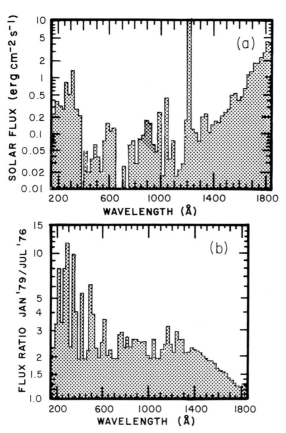

Fig. 6.6 a, b. Solar EUV fluxes on 22 January 1979 (**a**) and their ratio to the July 1976 fluxes (**b**). Measurements done with Extreme Ultraviolet Spectrometer of the Atmospheric Explorer Satellite. (After Hinteregger 1981)

ratios measured by different instruments, such is not the case with the absolute fluxes. This is particularly true at H Ly-α, He 584, He 304, and X-ray wavelengths. One must be cognizant of this reality when constructing aeronomical models, and choose the fluxes with caution.

The detailed ionospheric scheme is considerably more complex than the simplified version illustrated in Fig. 5.1. This is because charge transfer reactions involving minor constituents begin to be important in the lower ionosphere, particularly below the homopause. The present understanding of the important chemical reactions in the ionospheres of all major planets is summarized in Table 6.5, and schematically represented in Fig. 6.7. EUV or electron impact-ionization of H_2 produces H_2^+ and H^+ ions (Table 6.5). The latter are produced also on direct ionization of the atmospheric H atoms. Direct production of the H^+ ions can become dominant only at high latitudes, where large abundance of H atoms could result from charge particle dissociation of H_2 (Atreya et al. 1979b; Yung and Strobel 1980). Hence, the dominant source of the H^+ ions is the dissociative ionization of H_2, except at the auroral latitudes. Helium ions are responsible for the production of numerous intermediate ions; their abundances, however, are small due to the relatively low mole fraction of helium. Finally, CH_3 radicals produced in the CH_4-photochemistry can be ionized by the large Lyman-α solar flux and result in a narrow ionization layer below the homopause (Prasad and Tan 1974; Atreya and Donahue 1975a, b). Nearly half of the solar EUV flux absorbed in the upper atmosphere of Jupiter produces photoelectrons. The photoelectrons in turn cause additional ionization. This is, however, a secondary process contributing less than 25% and 5%, respectively, to the total column production of H_2^+ and H^+ (Waite et al. 1983).

In the ionospheres of the major planets, H_2^+ ions have the highest production rate, comprising between 90 and 95% of all ions produced. Their eventual ionospheric concentration is, however, insignificant as they undergo rapid charge transfer reactions with H_2 and H. At very high altitudes, where $[H] > [H_2]$, the following reaction is important

$$H_2^+ + H \rightarrow H^+ + H_2 \ .$$

Since the concentrations of H_2 and the hydrocarbons are low at these altitudes, H^+ is expected to remain as the terminal ion. As discussed earlier, H^+ ions are long lived, as they are slowly removed by radiative recombination, or by diffusion to the lower atmosphere where 3-body recombination loss becomes important (reaction $e10$ in Table 6.5). Under certain circumstances, when H_2 is vibrationally excited, the reaction of H^+ with H_2 ($v' \geq 4$), (reaction $e11$ in Table 6.5), could be important. In the ground vibrational state, the above reaction is endothermic by 1.8 eV at 300 K (Browning and Fryar 1973). In the middle atmosphere, where the H_2 densities are relatively large, the principal sink of H_2^+ ions is the following reaction ($e1$ in Table 6.5)

$$H_2^+ + H_2 \rightarrow H_3^+ + H \ .$$

The dissociative recombination of H_3^+ ions ($r1$, Table 6.5) is fast. At lower partial pressures (~ 0.01 mb) and temperatures (~ 205 K), the dimer H_5^+ (i.e., $H_3^+ \cdot H_2$) may be formed, instead of H_3^+. The recombination rate of H_5^+ is 3.6×10^{-6} cm^3 s^{-1} (Leu et al. 1973). Finally, the reactions of H^+ and H_3^+ ions with the hydrocarbons become important in the lower ionospheres where the hydrocarbon densities are appreciable. The terminal hydrocarbon ions are perhaps different from CH_5^+ and $C_2H_5^+$ (Table 6.5), since these ions can react rather rapidly with C_2H_2, C_2H_4 and C_2H_6 to form even heavier hydrocarbon ions (Strobel and Atreya 1983) in the following manner.

$$CH_5^+ + C_2H_2 \longrightarrow C_3H_5^+ + H_2$$
$$+ C_2H_4 \longrightarrow C_2H_5^+ + CH_4$$

(Munson and Field 1969) and

$$C_2H_5^+ + C_2H_2 \longrightarrow C_3H_3^+ + CH_4$$
$$\longrightarrow C_4H_5^+ + H_2$$

Table 6.5. Important chemical reactions in the ionospheres of the major planets[a]

Reaction number	Reaction	Rate constant	Reference
Ion production			
$p1$	$H_2 + h\nu(\lambda < 804 \text{ Å}) \rightarrow H_2^+ + e^-$		
$p2$	$\rightarrow H^+ + H + e^-$		
$p3$	$H_2 + e^- \rightarrow H_2^+ + 2e^-$		
$p4$	$\rightarrow H^+ + H + 2e^-$		1
$p5$	$H + h\nu(\lambda < 912 \text{ Å}) \rightarrow H^+ + e^-$		
$p6$	$H + e^- \rightarrow H^+ + 2e^-$		
$p7$	$He + h\nu(\lambda < 504 \text{ Å}) \rightarrow He^+ + e^-$		
$p8$	$He + e^- \rightarrow He^+ + 2e^-$		
$p9$	$CH_3 + h\nu(\lambda < 1262 \text{ Å}) \rightarrow CH_3^+ + e^-$		2
Charge exchange			
$e1$	$H_2^+ + H_2 \rightarrow H_3^+ + H$	2.0×10^{-9}	3
$e2$	$H_2^+ + H \rightarrow H^+ + H_2$	6.4×10^{-10}	4
$e3$	$He^+ + H_2 \rightarrow H_2^+ + He$	$\leq 2.0 \times 10^{-14}$ ⎫ sum	
$e4$	$\rightarrow HeH^+ + H$	$\leq 2.0 \times 10^{-14}$ ⎬ 1×10^{-13}	5
$e5$	$\rightarrow H^+ + H + He$	$\leq 8.0 \times 10^{-14}$ ⎭	
$e6$	$He^+ + CH_4 \rightarrow CH^+ + H_2 + H + He$	2.4×10^{-10}	6
$e7$	$\rightarrow CH_2^+ + H_2 + He$	9.3×10^{-10}	6
$e8$	$\rightarrow CH_3^+ + H + He$	9.6×10^{-11}	7
$e9$	$\rightarrow CH_4^+ + He$	1.6×10^{-11}	7
$e10$	$H^+ + H_2 + H_2 \rightarrow H_3^+ + H_2$	3.2×10^{-29}	8
$e11$	$H^+ + H_2(v' \geq 4) \rightarrow H_2^+ + H$	$k = fn(T_v)$	9
$e12$	$H^+ + CH_4 \rightarrow CH_3^+ + H_2$	2.3×10^{-9}	6
$e13$	$\rightarrow CH_4^+ + H$	1.5×10^{-9}	6
$e14$	$HeH^+ + H_2 \rightarrow H_3^+ + He$	1.85×10^{-9}	3
$e15$	$H_3^+ + CH_4 \rightarrow CH_5^+ + H_2$	2.4×10^{-9}	6
$e16$	$CH^+ + H_2 \rightarrow CH_2^+ + H$	1.0×10^{-9}	6
$e17$	$CH_2^+ + H_2 \rightarrow CH_3^+ + H$	1.6×10^{-9}	10
$e18$	$CH_3^+ + CH_4 \rightarrow C_2H_5^+ + H_2$	1.2×10^{-9}	10
$e19$	$CH_4^+ + CH_4 \rightarrow CH_5^+ + CH_3$	3.3×10^{-11}	10
$e20$	$CH_4^+ + H_2 \rightarrow CH_5^+ + H$	1.5×10^{-9}	10
Electron-Ion Recombination			
$r1$	$H_3^+ + e^- \rightarrow H_2 + H$, or	$2.8 \times 10^{-7} (200/T_e)^{0.7}$	11
	$\rightarrow H + H + H$		
$r2$	$H_2^+ + e^- \rightarrow H + H$	$< 1.0 \times 10^{-8}$	12
$r3$	$HeH^+ + e^- \rightarrow He + H$	$\sim 1.0 \times 10^{-8}$	13
$r4$	$H^+ + e^- \rightarrow H + h\nu$	4.0×10^{-12} $(250/T_e)^{0.7}$	14
$r5$	$He^+ + e^- \rightarrow He + h\nu$	4.0×10^{-12} $(250/T_e)^{0.7}$	14
$r6$	$CH_5^+ + e^- \rightarrow CH_4 + H$	3.9×10^{-6}	15
$r7$	$C_2H_5^+ + e^- \rightarrow C_2H_2 + H + H_2$	3.9×10^{-6}	15

[a] Adapted from Atreya and Donahue (1976); and Atreya et al. (1979b).
[1] McElroy (1973); [2] Atreya and Donahue (1975b); Prasad and Tan (1974); [3] Theard and Huntress (1974); [4] Karpas et al. (1979); [5] Johnsen and Biondi (1974); [6] Huntress (1974); [7] Adams and Smith (1976); [8] Miller et al. (1968); [9] Atreya et al. (1979b); Atreya and Waite (1981); McConnell et al. (1982); [10] Smith and Adams (1977); [11] Leu et al. (1973); [12] Auerbach et al. (1977); [13] Hunten (1969); [14] Bates and Dalgarno (1962); [15] Maier and Fessenden (1975).

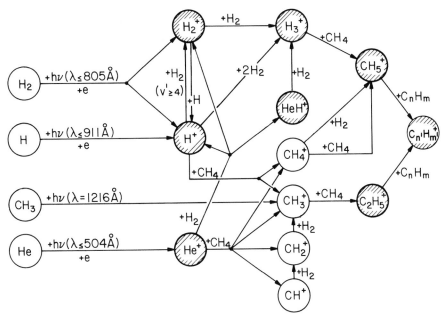

Fig. 6.7. Ionospheric chemistry on the major planets

or

$$C_2H_5^+ + C_2H_4 \longrightarrow C_3H_5^+ + CH_4$$
$$+ C_2H_6 \longrightarrow C_4H_9^+ + H_2$$

(Huntress 1977), followed by recombination reactions

$$C_nH_m^+ + e^- \longrightarrow C_2H_2, C_2H_4, C_2H_6 \ldots .$$

Laboratory chemical kinetics data for most of the above reactions, especially under Jovian conditions, is unavailable.

6.2.5 Model Ionospheres

Once the ion production and loss terms have been calculated, the ion or electron distribution can be obtained by solving the continuity equation, as discussed earlier. These equations are set up for each ion. They form sets of coupled partial differential equations which can be treated using standard numerical techniques. The column ion production rates calculated for the Voyager/Jupiter epoch are presented in Table 6.6. As expected, secondary ionization due to the photoelectrons is a small fraction of the total nonauroral ion production (Table 6.6). Also note the large production of the hydrogen atoms, hence the large direct ionization of H at the auroral latitudes. The

Table 6.6. Ion production rates at Jupiter[a]

	Production rates $[10^8 \text{ cm}^{-2}\text{ s}^{-1}]$		
	from EUV	from photoelectrons	from auroral particles
H_2^+	7.4	2.4	565
H^+			
from H_2	0.8	0.05	29
from H	0.7	0.04	870
He^+	0.035	0.0002	0.004
H	4	10	475

Total solar EUV absorbed on top of the atmosphere, $3 \times 10^{10} \text{ eV cm}^{-2}\text{ s}^{-1}$.
Total photoelectron energy, $1.8 \times 10^{10} \text{ eV cm}^{-2}\text{ s}^{-1}$.
Total auroral energy input, $6.3 \times 10^{12} \text{ eV cm}^{-2}\text{ s}^{-1}$ ($\sim 10 \text{ erg cm}^{-2}\text{ s}^{-1}$).
[a] Adapted from Waite et al. (1983). Entries in the "auroral" column are based on a beam of 1 keV electrons. A 10 keV beam with the same energy would give nearly two to four times the production rates listed here.

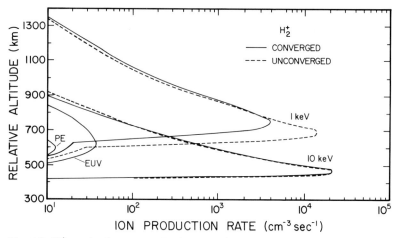

Fig. 6.8. H_2^+ production rates at Jupiter assuming photoelectron (PE), solar EUV, and 1 and 10 keV electrons of energy 10 erg cm^{-2} s^{-1} as the ionization sources. In the "unconverged" case the charged particles act on an atmosphere whose H and N_e concentrations are determined by EUV. The charged particles modify both H and N_e. The process is iterated until convergence is reached. Altitudes are above the 1-bar level. (Waite et al. 1983)

assumed auroral energy input is consistent with the Voyager observations. Further discussion of the auroral precipitation is deferred to a latter section on atmospheric heating. The production rates of H^+, H_2^+ as a function of altitude are shown in Figs. 6.8 and 6.9.

The resulting electron concentration profiles for the EUV and charged particle-dominated ionospheres are shown in Fig. 6.10. For steady state, a

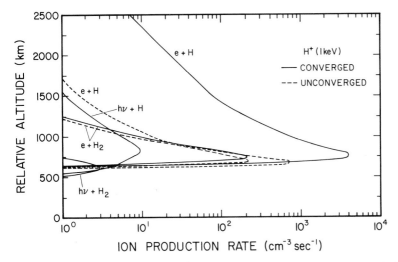

Fig. 6.9. Same as Fig. 6.8 except for H^+. Both EUV and the 1 keV electron sources include direct ionization of H, and dissociative ionization from H_2. (Waite et al. 1983)

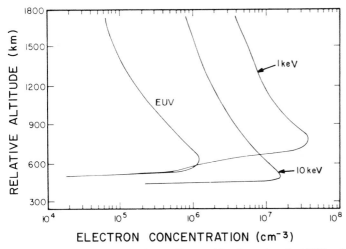

Fig. 6.10. Jovian electron density profiles resulting from the solar EUV and the charged particle sources. (Waite et al. 1983)

good estimate of the peak electron concentration may be obtained by simply equating H^+ production and loss rates, since chemical equilibrium prevails at these altitudes. Thus, at the electron peak,

$$P_H^+ = L_H^+ \equiv \alpha_r [N_e][H^+] \equiv \alpha_r [N_e]^2 \ , \tag{6.33}$$

since $[H^+] \approx [N_e]$ at the electron peak.

Taking $P_H^+ \approx 10\,\mathrm{cm}^{-3}\,\mathrm{s}^{-1}$ for EUV ionization, (Fig. 6.9), and α_r from Table 6.5 with $T \approx 200\,\mathrm{K}$ at the electron peak ($\sim 600\,\mathrm{km}$), one obtains $[N_e] \approx 1.5 \times 10^6\,\mathrm{cm}^{-3}$, which agrees well with the results of detailed calculations shown in Fig. 6.10. A quick comparison of the model calculations with the Voyager data (Fig. 6.2) reveals that there is at least a factor of 10 discrepancy between the two. Similar discrepancies exist also for the Saturn ionosphere.

6.2.6 Reconciliation of Models with Measurements

The above-mentioned disagreement between the calculated and the measured electron profiles could imply that some loss mechanism other than the slow recombination of the terminal ions, H^+, must be operative. Amongst the many possibilities is conversion of H^+ to some molecular ion, since molecular ions have relatively short lifetimes against recombination. A candidate for the needed conversion was alluded to earlier, i.e., a reaction between H^+ and H_2 ($v' \geq 4$). Some credence to this mechanism is provided by studying the comparison between model calculations and the Pioneer 10 ionospheric measurements (Fig. 6.11). The calculations were done in the same manner as for the Voyager epoch, and using the appropriate solar fluxes, ambipolar diffusion was included. For the most part, there seems to be a general agreement between the model and the Pioneer data. (The sharp layers in the lower ionosphere are reminiscent of the enhancements seen in the E-region of the Earth's ionosphere. Such enhancements are believed to be caused by long-lived metallic ions which tend to form layers due to an electric field resulting from wind shear. On Jupiter, Na^+ of the Io-origin may be a possible candidate, as proposed by Atreya et al. (1974) in their pre-Pioneer work. In addition, the ablation of meteorites in the atmosphere could potentially be a source of metallic ions for all of the outer planets. Information on the metallic ions and the upper atmospheric winds on the outer planets is lacking.) The only major difference between the Pioneer and the Voyager epochs is the considerably higher exospheric temperatures measured during the latter period. Indeed, higher temperatures would result in larger population of H_2 in the vibrationally excited states. In other words, at the Pioneer epoch, the loss of H^+ on H_2 ($v' \geq 4$) was perhaps not significant, and is also not required for a reasonably good fit of model calculations to the measurements. At the Voyager epoch, on the other hand, the above-mentioned loss mechanism was perhaps important, and it could in principle help reconcile the models with the data. A rate coefficient of $4 \times 10^{-9}\,\mathrm{cm}^3\,\mathrm{s}^{-1}$ would be needed for reaction $e11$ (Table 6.5) to reduce the Jovian electron concentration by a factor of 10 (Atreya et al. 1979b). Laboratory measurements of this rate constant are not available. It is, however, found that temperatures considerably warmer than those prevailing during the Voyager encounters would be needed to pump required quantities of H_2 into vibrationally excited levels (Atreya et al. 1979b; McConnell et al. 1982; Waite et al. 1983). Thus, the H^+ loss on H_2 ($v' \geq 4$) holds

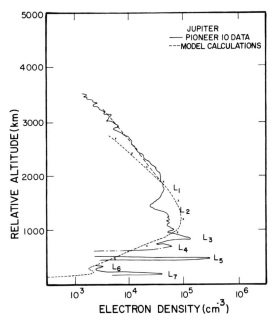

Fig. 6.11. Comparison of the Pioneer 10 electron density profile with model calculations. The geometry is: zenographic latitude of 26°N, solar zenith angle of 81°. The models are with radiative recombination coefficient (α_r) varying as $T_e^{-0.5}$: *short dashed curve*; also including diffusive equilibrium: *long dashed curve*. *Crosses* represent the case with $\alpha_r \propto T_e^{-0.75}$ but without diffusive equilibrium. (Atreya and Donahue 1976)

most promise for the auroral region where auroral energy input can result in large vibrational temperatures (McElroy 1973; Cravens 1974). Nearly 10% of all auroral energy ends up in vibrational excitation of H_2 (Waite et al. 1983).

Figure 6.12 illustrates an attempt at reconciling the model calculations with the Voyager ionospheric measurements at middle and auroral latitudes of Jupiter, taking into account the above-mentioned vibrational loss process. The vibrational temperature is parameterized. It is clear that high vibrational temperatures, around 3000 K, are needed for a reasonable agreement between the model results and the data. Only in the auroral region is there some possibility of encountering such high temperatures. Similar calculations for Saturn, including the H^+ sink on H_2 ($v' \geq 4$), are presented in Figs. 6.13 and 6.14. Again, it is found that only at the auroral latitudes does the likelihood of reconciling calculations with the measurement exist. Even then the agreement is at best marginal for Saturn.

Another candidate for converting H^+ ions to a molecular ion is H_2O. The possibility of atmospheric and ionospheric interaction with the rings has been alluded to by several authors (see, e.g., Atreya and Donahue 1975a; Shimizu 1980; Chen 1983; Ip 1983; Atreya 1984b; Atreya et al. 1984; Connerney and Waite 1984). Presence of water vapor at ionospheric heights would ensure conversion of H^+ to molecular ions H_2O^+ and H_3O^+ according to a scheme presented in Table 6.7.

The importance of the chemical sink given in Table 6.5 rests upon the available flux of water in the upper atmosphere. As noted in Chapter 3 on Cloud Structure, water intrinsic to the major planets is locked up in the water

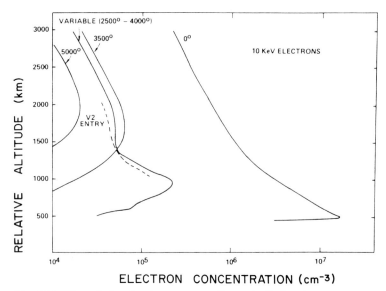

Fig. 6.12. Effect of vibrational temperature, T_v, on the Jovian electron concentrations. At high T_v, particularly around 2500 to 4000 K, the H^+ loss on H_2 ($v' \geq 4$) becomes significant. A reasonable agreement between the model and the measured N_e at high latitudes is possible under these conditions. (Waite et al. 1983)

clouds deep in the troposphere. Any upper atmospheric water therefore must have an extraplanetary origin. At Saturn, perhaps the rings can supply the needed flux of 4×10^7 H_2O molecules $cm^{-2}s^{-1}$, planetwide average. In this regard it is interesting to note that the Voyager/UVS observations indicate a Lyman-α emission of $350\,R$ associated with the B rings, as shown in Fig. 6.15. This would imply an H column abundance of $10^{13}\,cm^{-2}$, or 600 H atoms cm^{-3} assuming a circular torus of 1 R_s (Saturn radius) extent (Broadfoot et al. 1981a). Removal of water from the rings by micrometeoroid impact (Morfill et al. 1983), and neutralization of ionospheric H^+ (Atreya et al. 1984) can both result in a thin H atmosphere around the B ring. The former, however, yields the largest amount of water flux (0.6 to 4.5×10^7 $cm^{-2}s^{-1}$), 10% of which is expected to be deposited into Saturn's atmosphere. At Jupiter, large H_2O flux from the Galilean moons is not expected. However, as discussed in the context of Jovian CO (Chap. 5 on Photochemistry), influx of O atoms or ions is possible, and subsequent chemical reactions can give rise to OH in the upper atmosphere. Another source of water molecules, of course is the micrometeoritic infall (Chap. 5). It is important to note that no reliable detection of any oxygen-bearing molecules in the upper atmospheres of the major planets is as yet available. In view of such lack of critical data, the H^+ sink on H_2O can be regarded as only potentially important.

Finally, the flux divergence term in the continuity equation [Eq. (6.1)] represents effectively an ionospheric loss. With the exception of ambipolar

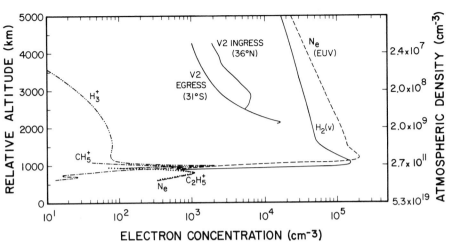

Fig. 6.13. Comparison of the Saturn electron concentration models assuming EUV and low latitude vibrational population of H_2, $H_2(v)$, with the Voyager low latitude data. Altitudes are above the 1-bar level. (Atreya et al. 1984)

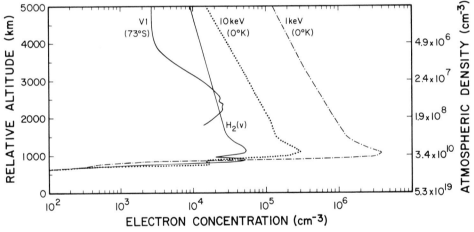

Fig. 6.14. Comparison of Saturn ionospheric models assuming 1 and 10 keV electrons and 1 keV electrons with high latitude population of vibrationally hot H_2, with a high latitude measurement done from Voyager 1. (Atreya et al. 1984)

diffusion, the ionospheric models thus far do not account for several of the plasma transport processes that could be important on the major planets. The effect of these processes on the ionospheric structure is to redistribute the ions. The ion peak could be elevated or lowered, there could be an apparent reduction in the local electron concentration, and the ionosphere could appear to be quite extensive. Indeed, if one takes the Voyager measurements on their

Table 6.7. H^+ Loss on H_2O

$$
\begin{aligned}
H_2O + h\nu &\rightarrow OH + H \\
H^+ + OH &\rightarrow OH^+ + H \\
OH^+ + H_2 &\rightarrow H_2O^+ + H \\
H_2O^+ + H_2 &\rightarrow H_3O^+ + H \\
H_3O^+ + nH_2O &\rightarrow H_3O^+ \ (H_2O)_n
\end{aligned}
$$
and
$$
\begin{aligned}
OH + H_2 &\rightarrow H_2O + H \\
H^+ + H_2O &\rightarrow H_2O^+ + H
\end{aligned}
$$
followed by
$$
H_2O^+, \ H_3O^+ \text{ or } H_3O^+ \ (H_2O)_n + e^- \rightarrow \text{neutrals}
$$

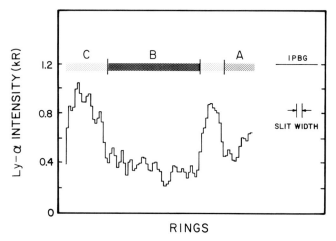

Fig. 6.15. Voyager UVS slit scan across the illuminated portion of Saturn's rings. The rise in signal between B and A rings is due to the transmission of the interplanetary background (IPBG) by the Cassini division. This contamination can be avoided if observations of the B ring are used since its Ly-α albedo is negligible and the slant optical depth for the observations greater than 3. (Broadfoot et al. 1981a)

face value, it would seem that the peak in the electron concentration is measured to be up to 1500 km higher than predicted by the models. Vertical drifts on the order of tens of $m\,s^{-1}$ would be needed to raise the peak by this amount. At Saturn midlatitudes, e.g., diurnal vertical drift velocity of $14\,m\,s^{-1}$ for H^+, driven by an electric field of $3\,mV\,m^{-1}$, southward wind of $140\,m\,s^{-1}$ and/or vertical winds of $14\,m\,s^{-1}$ are required to raise the calculated peak in the electron profile to its observed altitude (Atreya and Waite 1981). At Jupiter, plasma drifts of $\sim 50\,m\,s^{-1}$ are not unlikely (Strobel and Atreya 1983). Although no direct measurements of the Jovian thermospheric dynamics are yet available, it is reasonable to expect that the enormous power dissipated into the auroral regions would result in heating of the upper atmosphere, which would subsequently drive a massive wind system. In the equatorial region, vertical transport across the horizontal magnetic field lines can be effected by dynamo electric fields and $(E \times B)$ drifts.

6.2.7 Uranus and Neptune Ionospheres

Since the south pole of Uranus was pointing to the Sun at the time of Voyager 2 observations, the radio occultation regions (2° and 6°) would have experienced a diurnal cycle. This is because of Uranus' equator-to-orbit inclination of 98° that results in perpetual daylight at latitudes greater than 8°S, perpetual darkness at latitudes greater than 8°N, and a diurnal cycle at latitudes between 8°S and 8°N. Of course, the situation would change as Uranus moves in its orbit around the Sun. For the Voyager epoch, then, the ionospheric models should assume diurnally averaged solar fluxes appropriate for the solar minimum conditions. The electrons responsible for the Uranus electroglow (see Chap. 4 on Vertical Mixing) do little by way of ionizing the major atmospheric constituents. Furthermore, there is evidence of only an auroral hot spot in the vicinity of the south magnetic pole, so that ionization due to any widespread auroral charged particle precipitation is out of the question. Thus, the ionospheric structure at Uranus is expected to be controlled principally by the solar EUV.

The pre-Voyager ionospheric models of Uranus by Atreya and Ponthieu (1983) and Atreya (1984a) indicated a peak electron concentration of 2×10^4 cm^{-3}, with a nominal value around 3000 to 5000 electrons cm^{-3}, when plasma transport and a high exospheric temperatures (250 – 300 K) are considered. The peak occurs at an altitude of ~ 1000 km (above 1 bar) where the atmospheric density is $\sim 3 \times 10^{10}$ cm^{-3}. The Voyager/UVS observations have shown that the exosphere is even hotter (750 K), with the consequence that Uranus possesses an extensive corona of H_2 and H (see Chap. 2 on Thermal Structure). An immediate implication for the ionosphere is that it, too, would be quite extensive, as is apparently the situation (see Sect. 6.1). Furthermore, direct ionization of atomic hydrogen should become a major contributor to the total electron concentration at altitudes greater than approximately 5000 km. A preliminary model of the Uranus ionosphere based on the present neutral atmospheric model (see Chap. 2 on Thermal Structure) is shown in Fig. 6.16. As before, the peak electron concentration is close to 1.5×10^4 cm^{-3}, with the most likely value in the neighborhood of 5000 electrons cm^{-3}. The major topside ion is H^+, which has a long lifetime ($\sim 10^8$ s). Indeed at large planetocentric distances in the magnetosphere of Uranus, H^+ is the only ion detected, unlike Jupiter and Saturn, where H_2^+ and H_3^+ (presumably of ionospheric origin) were also detected. It is important to recognize that, despite great differences between the neutral atmospheric models, the electron concentration profile of Fig. 6.16 is virtually the same as the pre-Voyager

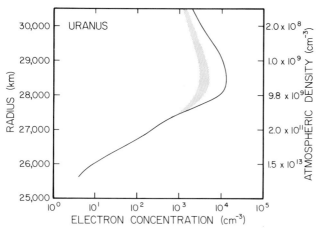

Fig. 6.16. A model of the ionosphere of Uranus in the region of the Voyager radio occultation experiment. *Stippled area* represents the "best guess" electron density model taking into account first-order plasma transport effects. Atmospheric densities corresponding to the planetocentric distances (radius) are given on the *right ordinate*

model of Atreya (1984a), provided that the comparison is made of the electron concentrations plotted against the atmospheric density (not altitude). The only major departure arises above the electron peak, where the topside scale height is now much greater than in the pre-Voyager model. As the spacecraft attitude information is refined, there may be some changes in the radius versus atmospheric density. This should not, however, result in any major changes in the electron concentration profile, if plotted against the atmospheric density. The layers detected in the ionosphere of Uranus could be due to the ionization of metals of meteoritic origin, as discussed earlier in the context of Jupiter. It is unlikely that migration to the ionosphere of large quantities of material from the rings or the moons will occur. Because of Uranus' relatively placid magnetospheric environment, erosion of the ices on the surfaces of the rings and the moons is expected to be minimal. Note. however, that charged particle erosion of the methane ice (which presumably covers the Uranus ring and moon surfaces) has been proposed as a mechanism to explain the dark appearance of the rings and moons of Uranus (L. J. Lanzerotti 1986, personal communication). Such an erosion will leave behind a carbon residue. Even if carbon could enter the atmosphere, it is safe to assume that it will not modify the composition or the structure of the upper atmosphere and ionosphere of Uranus.

    At the auroral latitudes, the Uranus ionosphere is expected to be controlled principally by the charged particles. There are no measurements available of either the ionospheric profiles or the charged particles at high latitudes of Uranus. Nevertheless, it is instructive to model the ionospheric structure at these latitudes, as it has important magnetospheric applications. A model of

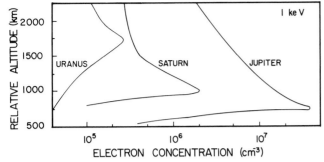

Fig. 6.17. Models of the high latitude ionospheres of Jupiter, Saturn and Uranus assuming 1 keV electrons and power inputs of respectively 10^{13}, 10^{11}, and 10^{11} W. Altitudes are above the 1-bar level

the Uranus electron profile at auroral latitudes along with its comparison with the auroral ionosphere models of Jupiter and Saturn is shown in Fig. 6.17. These profiles would indicate large ionospheric conductivities on all three planets (see Sect. 6.3.2 on Joule Heating). Information about the magnetospheric environment of Neptune is expected to be available from the Voyager observations in 1989. The EUV-dominated ionosphere, however, is predicted to have a peak electron concentration of approximately $1000 - 3000$ cm^{-3} (Atreya 1984a).

6.3 Upper Atmospheric Heating

As discussed in Chap. 2 on Thermal Structure, the thermospheres of Jupiter, Saturn, and Uranus are much warmer than the solar EUV and the internal heat would allow. For example, Jupiter's exospheric temperature is about 1000 K, rather than the suspected value of $150 - 200$ K. No data on the thermospheric dynamics of these planets is available, therefore the role of convection in heat transport is merely speculative. Conduction, on the other hand, is expected to be the principal mode for transporting heat, as in the terrestrial thermosphere. A first-order approximation of the thermospheric temperature profile can be obtained by solving the one-dimensional heat conduction equation, which has the following form

$$F = \kappa \frac{dT}{dz} , \qquad (6.34)$$

where κ is the atmospheric conductivity, and F is the heat flux. κ can be further expressed in the following empirical form

$$\kappa = A T^S \ [\text{erg cm}^{-1} \text{s}^{-1} \text{K}^{-1}] , \qquad (6.35)$$

where $A = 252$, and $s = 0.751$ for H_2 at the high exospheric temperatures of Jupiter (Hanley et al. 1970). Also, the heat flux can be written in terms of its sources and sinks, q and l, i.e.,

$$F = \int (q-1)\,dz \ . \tag{6.36}$$

Substitution of Eq. (6.35) into Eq. (6.34) and subsequent integration yields the following expression for temperature at any height.

$$T(z) = T_m + \{(s+1)F\Delta z/A\}^{1/(s+1)} \ . \tag{6.37}$$

In the above expression, Δz is the interval between the heat source and the heat sink levels, assuming that they are uniquely defined, and T_m is the temperature at the lower reference level.

On the Jovian planets, the sources of thermospheric heat are solar EUV, soft electrons, energetic (auroral) electrons or other charged particles, and potentially Joule heating and inertia gravity waves. The heat is conducted down into the lower atmosphere where infrared cooling by the hydrocarbons occurs, since H_2 is a poor infrared radiator.

6.3.1 Heat Sink

Both CH_4 and C_2H_2 are efficient radiators in the infrared. Since the C_2H_2 abundance is $\sim 10^{-4}$ of the CH_4 abundance in the Jovian and Saturnian stratospheres, the former competes with CH_4 in infrared cooling only near the homopause where the two can have comparable abundance. The radiative cooling rates, R_i, can be calculated in the following manner (Strobel and Smith 1973)

$$R_i = h\,v_i\,A_i\,n_i^* \ , \tag{6.38}$$

where v_i and A_i are, respectively, frequency and probability of $v = 1$ to $v = 0$ transition, n_i^* is the population density in the $v = 1$ level of the constituent doing the IR cooling.

Using Boltzmann distribution for n_i^*, the following expression for R_i is obtained

$$R_i = \frac{h\,v_i\,n_i \sum\limits_j P_{1\to0}^{ij} Z_{ij}\,n_j \dfrac{\omega_1}{\omega_0} \exp(-h\,v_i/kT)}{A_i + \sum\limits_j P_{1\to0}^{ij} Z_{ij}\,n_j} \ , \tag{6.39}$$

where Z_{ij} is the elastic (hard sphere) collision frequency between molecules i and j, n_j is the density of the jth collision partner, $P_{1\to0}^{ij}$ is the collisional deactivation probability for collisions between i and j molecules, ω_0 and ω_1 are statistical weights of levels $v = 0$ and $v = 1$, respectively. The collisional deactivation terms are poorly known. At Jupiter, the column heating rate is balanced by the above-mentioned column cooling rate at around a microbar level. On Saturn, this occurs at a lower pressure of about a nanobar, since the hydrocarbon densities are appreciable even at much lower pressures compared to Jupiter because of Saturn's larger vertical eddy diffusion coefficient.

6.3.2 Heat Sources

The deposition of energy by the solar EUV and the charged particles results in the formation of ionosphere, excitation, and emission of aurora and airglow features, and atmospheric heating. These energy sources, and some others relevant to thermospheric heating, are discussed below.

EUV *Heating*

Typical energies of the solar photons responsible for the above-mentioned processes lie between 10 and 100 eV (1000 to 100 Å). The ionization potential of the major atmospheric constituent, H_2, is 15.41 eV. Thus, a large portion of the incoming EUV flux has energy far in excess of the ionization threshold. The photoelectrons generated in the ionization process carry the excess energy. As mentioned earlier, they in turn continue the ionization process until their energy falls below the ionization potential. In general, the energy associated with the photoelectrons, E_{PE}, is given by the following expression.

$$E_{PE} = h\nu - E_{IP} \ . \tag{6.40}$$

When $E_{PE} < E_{IP}$, the photoelectrons lose their energy in elastic and inelastic collisions with ambient neutral and plasma constituents. In inelastic collisions, kinetic energy is not conserved and the photoelectrons lose their energy by excitation of the neutrals at the following rate

$$\frac{dE_{PE}}{dx} = n E_{ex} \sigma_{ex} \ , \tag{6.41}$$

where n is the number density of the constituent, σ_{ex} is its excitation cross-section, and E_{ex} is the excitation energy. On Jupiter, approximately 10% of the photoelectron energy results in the H_2 Lyman and Werner band excitations (Waite et al. 1983).

At relatively low energies, elastic Coulomb collisions – i.e., collisions in which the kinetic energy is conserved – of photoelectrons occur most efficiently with the ambient electrons. This can cause electron temperatures to be substantially elevated above the ion and neutral temperatures.

Thermal equilibrium between electron, ion, and neutral temperatures is expected to be maintained up to altitudes well above the electron peak on the major planets (Henry and McElroy 1969; Nagy et al. 1976). Below the peak, electrons cool efficiently through rotational and vibrational excitation of H_2. Above the peak, they lose their energy to the ambient ions, H^+. The H^+ ions in turn cool by elastic charge exchange with the neutral gas, and by resonance charge exchange with H atoms.

The efficiency of direct heating of the neutral gas by photoelectrons is close to 15% at Jupiter. The major EUV heat source at the outer planets is actually chemical heating available as a result of H_2 ionization and its subsequent dissociation, as outlined below.

$$H_2 + h\nu \longrightarrow H_2^+ + e^-$$

followed by

$$H_2^+ + H_2 \longrightarrow H_3^+ + H$$

and

$$H_3^+ + e^- \longrightarrow H_2 + H$$
$$\xrightarrow{\text{or}} H + H + H \ .$$

In the above process, 15.41 eV is needed to ionize one H_2 molecule, and 4.479 eV is needed to dissociate it, so that 10.931 is available per H_2 ionization for heating of the neutral gas. In the topside, where $[H] > [H_2]$, H_2^+ would preferentially charge exchange with H, so that the above-mentioned sequence of reactions becomes invalid, and little, if any, chemical heating is possible. Despite the relatively large chemical heat source (except very high in the atmosphere), it is important to note that the total column integrated heat absorbed is a meagre $1.3 \times 10^{-2}\,\mathrm{erg\,cm^{-2}s^{-1}}$ at Jupiter, $3 \times 10^{-3}\,\mathrm{erg\,cm^{-2}s^{-1}}$ at Saturn, $10^{-3}\,\mathrm{erg\,cm^{-2}s^{-1}}$ at Uranus and $4 \times 10^{-4}\,\mathrm{erg\,cm^{-2}s^{-1}}$ at Neptune. This means that even at Jupiter the total increase in the temperature above its homopause value, due to the direct photoelectron and chemical heat sources, is less than 100 K, i.e., about 800 K less than the measured value.

Charged Particle Impact

Magnetospheric soft electrons and energetic charged particles result in atmospheric heating in the same manner as photoelectrons. Nearly half of the auroral energy input ends up in heating the neutral gas. Unlike photoelectrons, however, these charged particles are mostly confined to high (auroral latitudes). The altitude of their energy deposition is a function of the energy spectrum of the incoming beam, with more energetic particles depositing their energy deeper in the atmosphere. Heavy ions, on the other hand, have a shorter range than the electrons with corresponding energy. The auroral energy deposition, in addition to causing local heating, may also result in planetwide atmospheric heating if efficient thermospheric wind circulation prevails. The observed airglow and auroral emission rates, and corresponding energy inputs on the major planets are summarized in Table 6.8.

The aurora at Jupiter appears to be modulated by Io. This is particularly evident in Fig. 6.18 prepared from the north-south scans of Jupiter by the Voyager/UVS slit. The equatorward boundary of the observed aurora in this figure roughly coincides with the magnetic mapping of the Io torus on to Jupiter's atmosphere. The poleward boundary of the aurora is not known

Table 6.8. Airglow and auroral emissions and energies

Energy/emission		Jupiter	Saturn	Uranus
EUV energy absorbed [erg cm^{-2} s^{-1}]		1.3×10^{-2}	3×10^{-3}	1×10^{-3}
Airglow (low-mid latitudes)				
He 584 Å (R)		~4	$2.2 - 4.2$	0.11 ± 0.08
Lyman-α (kR)	Day	14	$2.5 - 3.0$	1.5
	Night	0.7	0.35	0.17
H$_2$ Lyman and Werner bands (kR)	Day	2.8 ± 1	0.7	0.4
	Night	<0.3	<0.01	Negligible
Energy input implied by the H$_2$ band intensities [erg cm^{-2} s^{-1}]		0.3	0.05	0.075 (from electro-glow)
Aurora				
Auroral zone		Magnetic mapping of Io-torus	$78° - 81.5°$	$15° \times 25°$ ellipse around the magnetic pole
Lyman-α (kR)		60	10	1.5
H$_2$ Lyman and Werner bands (kR)		$60 - 80$	$10 - 15$ (sporadically up to $100\,kR$	9
Hemispherical auroral power input [W]		1.3×10^{13}	2×10^{11}	1×10^{11}
Energy flux into auroral zone [erg cm^{-2} s^{-1}]		10	0.7	2
Globally averaged auroral energy input [erg cm^{-2} s^{-1}]		0.4	0.01	0.025

precisely because of some uncertainty in the position of the slit on Jupiter. This problem is aggravated by the fact the slit size is relatively large $(0.1° \times 1°)$ and the auroral emissions occur near the limb at high latitudes. Although the Jovian auroral excitation can be interpreted in terms of the energetic electron precipitation, excitation due to heavy ions, such as sulfur and oxygen ions from the Io-torus, is equally likely (Gehrels and Stone 1983). The aurora on Saturn is basically a magnetotail type aurora, and as on the Earth, it is driven by the solar wind charged particles.

It is evident from the low latitude H$_2$ airglow and the implied energy input (Table 6.8) that some energy source other than the EUV must be present. It is further confirmed by the fact that both Lyman-α and the H$_2$-bands have non-negligible nightside emissions, which cannot of course be due to photoelectrons. On Jupiter, if the auroral energy could be spread uniformly over the entire planet with 100% efficiency, e.g., by the thermospheric winds, the

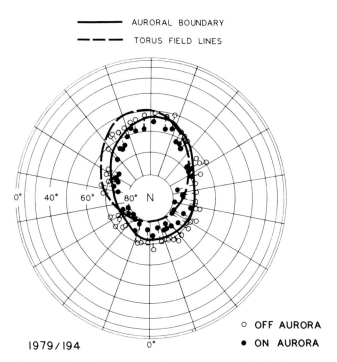

Fig. 6.18. Voyager UVS tracks of the Jovian aurora. *Dashed curve* shows magnetic mapping of Io plasma torus on to Jupiter. *Solid line* gives equatorward boundary of the auroral zone inferred from the UVS observations. *Open circles* represent the positions of the northern end of the UVS slit when no aurora was recorded; *solid circles* give its positions when an aurora was detected. (Broadfoot et al. 1981b)

planetwide average energy input (due to the auroras) would be 0.4 erg cm^{-2} s^{-1} (Table 6.8). This is of the same order as needed to excite the observed equatorial H_2-bands. At this time, nothing is known about the thermospheric winds on the outer planets. On Saturn, even if the auroral energy could be distributed over the entire planet, there would still be an energy crisis, since nearly five times greater energy is needed to excite the observed H_2-bands in the equatorial region (Table 6.8). Precipitation of magnetospheric soft electrons has been proposed as a possible mechanism for explaining both the non-auroral H_2-band intensities and the high exospheric temperatures (Hunten and Dessler 1977).

An example of the thermospheric heating resulting from input of the various above-mentioned heat sources on Jupiter is shown in Fig. 6.19. The 1 and 10 keV auroral electrons with energy flux of ~10 erg cm^{-2} s^{-1} result in the exospheric temperatures in the auroral region of the order of 2000 to 3000 K. Joule heating (next section), which is expected to be important at high

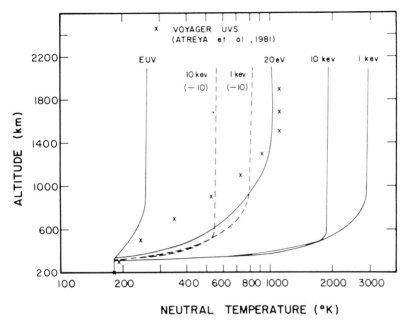

Fig. 6.19. Jovian neutral temperature models assuming various heat sources: photoelectrons (EUV), soft magnetospheric electrons (20 eV) with 0.5 erg cm^{-2} s^{-1} energy, energetic auroral electrons (1 and 10 keV) with 10 erg cm^{-2} s^{-1} energy, and energetic electrons with their energy spread over the planet ($\div 10$). Also shown with *crosses* are the Voyager measurements of the temperature. (Waite et al. 1983)

latitudes, would raise these temperatures even more. Strong thermospheric winds, on the other hand, would tend to spread the auroral energy over the entire planet. In Fig. 6.19 this effect is roughly simulated by diluting the auroral energy by a factor of 10 (approximate ratio of auroral to planetary hemisphere area). As seen in this figure, the 1 keV result with hemispheric average gives thermospheric temperatures quite close to those measured, it does not, however, reproduce them entirely. The match with the data is quite acceptable, if the heating were being done by 20 eV electrons carrying 0.5 erg cm^{-1} s^{-1} energy. Thus, it seems that some combination of the auroral and soft electron energy source would be needed to satisfactorily explain the thermospheric temperatures and the H$_2$-band intensities at low to mid latitudes of Jupiter. On Saturn, the situation is rather grave, since approximately 0.1 erg cm^{-2} s^{-1} would be needed to explain the 400 K exospheric temperature, whereas a maximum of only half this energy input can be presently accounted for. And, if the exospheric temperature were closer to 800 K, then approximately 0.3 erg cm^{-2} s^{-1} of energy input would be necessary. The soft electron hypothesis could be even more attractive for Saturn. Some mechanism must still be found, however, to precipitate the magnetospheric electrons in the equatorial and midlatitude regions.

At Uranus, the energy flux required to explain the observed exospheric temperature of 750 K is on the order of 0.1 to 0.15 erg cm^{-2}s^{-1}. The electroglow implies a planetwide energy input which falls short by a factor of 2 to 3 (additional details on Uranus electroglow are contained in Chap. 4 on Vertical Mixing). The electroglow energy can, however, be supplemented by the auroral energy (Table 6.8) provided that the latter is distributed over the entire planet, e.g., by the thermospheric winds. Electroglow and the associated energy input are expected to be present at Saturn as well, although to a lesser degree than at Uranus.

Joule Heating

Joule or frictional heating results from the presence of electric currents in the ionospheres of planets with magnetic fields. The dissipation of electrical energy (called Joule dissipation) occurs through charged particle collisions, giving rise to neutral and plasma heating. In its most rudimentary form, Joule heating rate, Q_J, is given by the following expression

$$Q_J = j \cdot E \ , \tag{6.42}$$

where j is the electric current and E is the electric field. Using Ohm's law, one can write

$$j = (\sigma)E \ , \tag{6.43}$$

where (σ) is the conductivity tensor. The Joule heating rate for the currents flowing perpendicular to the magnetic field B is given by

$$Q_P = \sigma_P E^2 = j/\sigma_C \ , \tag{6.44}$$

where σ_P is called Pedersen conductivity, and σ_C is Cowling conductivity. The latter is large close to the magnetic equator. Large ionospheric currents, called the equatorial electrojet, in the Earth's ionosphere result from Cowling conductivity. The Cowling conductivity is related to both Pedersen as well as Hall conductivity by the following expression

$$\sigma_C = \frac{\sigma_P^2 + \sigma_H^2}{\sigma_P} \ , \tag{6.45}$$

where σ_H is Hall conductivity which is perpendicular to both the electric and the magnetic fields. The general expressions for σ_P, σ_H and σ_L (Longitudinal conductivity – that parallel to B) conductivities for upper ionospheres that are dominated by a single ion are given below

$$\sigma_L = \frac{N_i e^2}{m_i v_{in}} + \frac{N_e e^2}{m_e v_{en}} \approx \frac{N_e e^2}{m_e v_{en}} \tag{6.46}$$

since $N_i \approx N_e$, and of course, $m_e \ll m_i$ and $m_e v_{en} \ll m_i v_{in}$. The same approximations are applied to σ_P and σ_H, so that

$$\sigma_P = \frac{N_e e^2}{m_i} \frac{v_{in}}{v_{in}^2 + f_i^2} \approx \frac{N_e e}{B} \frac{v_{in} f_i}{v_{in}^2 + f_i^2} \tag{6.47}$$

$$\sigma_H = \frac{N_e e^2}{m_i} \frac{f_i}{v_{in}^2 + f_i^2} , \tag{6.48}$$

where N_e represents the electron (or ion) concentration, m_e and m_i are, respectively electron and ion masses, v_{en} and v_{in} are, respectively electron-neutral and ion-neutral collision frequencies, f_i is the ion cyclotron frequency ($f_i = eB/m_i$).

Using the above expressions and the electron concentrations from the model ionospheres discussed earlier, the height-integrated Pedersen conductivities are found to be, at Jupiter: 0.2 mho[2] (equatorial) and 10 mho (auroral), (Strobel and Atreya 1983), and at Saturn: 5 mho (equatorial), and 60 mho (auroral) (Atreya et al. 1984). These estimates, however, are perhaps a factor of 10 to 50 too high, particularly for low latitudes where model ionospheric densities are much greater than those actually measured. The conductivity at Jupiter, however, should be comparable to that in the Earth's ionosphere, since although the ion gyrofrequency at Jupiter is 50 to 200 times greater, the electric fields are also expected to be much stronger than at the Earth. One estimate based on the observed departure of magnetospheric plasma from planetary corotation (McNutt et al. 1979; McNutt 1983) places the Jovian electric field at around 3 V m^{-1} (Waite 1981). This would imply an energy input of ~ 5 erg cm^{-2} s^{-1} at Jupiter's auroral latitudes (detailed calculations by Nishida and Watanabe 1981 indicate ~ 10 erg cm^{-2} s^{-1} of Joule heating input). At Saturn, nearly 10% departure of magnetospheric plasma from planetary corotation was observed (McNutt 1983). This would imply an energy input rate of ~ 0.15 erg cm^{-2} s^{-1}. Thus at the auroral latitudes of both Jupiter and Saturn, Joule heating rate appears to be competitive with the auroral heating. The Joule heating mechanism is potentially a very significant one for all of the major planets. Its full impact can be evaluated only when data on the thermospheric winds and electric fields become available. Conversely, one can obtain an idea of the thermospheric winds by using the following expression of Joule heating for the earth analog (Rees and Roble 1975).

$$Q_J = n \mu_{in} v_{in} \bar{W}^2 , \tag{6.49}$$

where $\bar{W} = V_n - V_i$ is the differential velocity between the neutrals and the ions. Calculations indicate that $\bar{W}$ on the order of 200 m s^{-1} would be needed to explain the 1000 K exospheric temperature at Jupiter (Atreya et al. 1981).

---

[2] 1 mho m^{-1} = 9×10^9 s^{-1} in Gaussian units.

Inertia Gravity Waves

The oscillations seen in the Jovian temperature profile derived from the ground-based stellar occultation observations (Chap. 2 on Thermal Structure) have been interpreted as vertically propagating gravity waves (French and Gierasch 1974). The absence of "flashes" or "spikes" in the light curves above the microbar level of those data would imply damping of these waves above this level. The energy carried by the gravity waves is estimated to be ~ 3 erg $cm^{-2} s^{-1}$ (French and Gierasch 1974). The waves generally deposit their energy 5 to 10 scale heights above the mesopause. If the dissipation of gravity waves manifests itself as heat, then a thermal gradient three to seven times greater than the observed value (Atreya et al. 1981) should exist. This fact makes the gravity wave mechanism less attractive for Jupiter. Nevertheless, it could be a potentially important contributor to another major heat source. Classical tidal wave calculations, along with the data on atmospheric dynamics, are needed for evaluating the importance of this source to thermospheric heating on the major planets.

6.4 Outstanding Issues

What is the structure of the lower ionospheres of Jupiter and Saturn? Where is the maximum in the electron concentration really located? What is responsible for the ledges (or sharp ionization layers) in the ionospheric data? What is the distribution of negative and cluster ions? What is the latitudinal and temporal behavior of the ionospheres? What is the composition of the ions? What is truly the abundance of H_2O and the vibrationally excited H_2 at ionospheric heights? Can the rings of Saturn and/or meteorites supply sufficient water to the upper atmosphere for the needed depletion of the ionosphere? Does the needed ion loss result from chemical processes, or is it simply the effect of ion transport? What is the magnitude of thermospheric winds and electric fields? Can energetic ions, such as sulfur and oxygen, explain the ionospheric structure as well as the auroral excitations seen at Jupiter? What is the energy spectrum of these ions and other charged particles, such as electrons and protons, on top of the atmosphere? Which of the many possible heat sources (Joule heating, charged particles, gravity waves, etc.) is responsible for the thermospheric heating on the outer planets? Repeated radio occultation observations from orbiters and in-situ measurements with ion mass spectrometers are necessary to answer these questions. They are necessary also for making significant headway in theoretical modeling of the ionospheres of the major planets.

7 Satellites

7.1 Atmospheres on Satellites

The presence of an atmosphere on planetary satellites is an anomaly. The relatively low masses of the moons result in low values of gravity and escape velocity. As a consequence, any volatiles outgassed during and after the accretionary phase of the moons would be rapidly lost. Since the conventional dynamo mechanism does not operate on the moons, they do not possess intrinsic magnetic fields. Therefore, they are likely to be stripped of any remaining atmosphere by interaction with charged particles of the solar wind or planetary magnetospheres origin. The only exceptions to the above scenario are: Saturn's moon Titan − which definitely has a massive atmosphere, Jupiter's moon Io − which appears to possess a tenuous atmosphere, and Neptune's moon Triton − which has some promise of having an atmosphere with pressures comparable to those in the troposphere of the Earth. In addition, extremely tenuous atmospheres might also exist on some of the Galilean moons (Europa, Ganymede, and Callisto), and above the rings of Saturn. Io's atmosphere seems to result from its perpetual volcanic activity caused by its unique tidal interaction in the Jovian system. The ability of Titan to retain an atmosphere appears to be the result of its comparatively greater mass, combined with its location in the Saturn system and in the primitive solar nebula. The reason for the existence of an atmosphere on Triton is similar to that for Titan, along with the fact that Triton should have strong tidal interaction with Neptune due to its 160° orbital inclination. The physical characteristics of the solar system satellites suspected of having an atmosphere (and their comparison with the Earth's moon) are presented in the General Appendix at the end of Chapter 7. Atmospheric and ionospheric characteristics and the related physics and chemistry of the individual moons are discussed in the following sections.

7.2 Io

In the year 1610, Galileo Galilei discovered Io, along with the other three large moons of Jupiter − Europa, Ganymede, and Callisto. These satellites were later named the Galilean moons of Jupiter, after their discoverer. The first clue to a possible atmosphere on Io was provided by a well-defined ionospheric structure measured in 1973 by the Pioneer 10 radio occultation experiment (Kliore et al. 1974), both on the dayside and nightside of Io. In the same year Brown (1974) detected Doppler-shifted D1 and D2 lines of sodium at 5890 and 5896 Å in a spectrum of Io. Potassium was subsequently detected in Io's environment by Trafton (1975). Kupo et al. (1976) reported discovering singly ionized sulfur in the magnetosphere of Jupiter. Just prior to the Voyager 1 encounter with Io in 1979, Peale et al. (1979) predicted that tidal heating should melt Io's interior, resulting in surface activity. The Voyager observations conclusively show that Io has exceptionally violent and persistent volcanic activity, which is expected to result in a bound atmosphere, and that it possesses an extended environment consisting of a host of neutral and ionized constituents. The discussion of Io in this chapter focuses on the atmospheric and ionospheric aspects, while magnetospheric plasma and volcanism are discussed only as they relate to our understanding of Io's aeronomy.

7.2.1 SO$_2$ Atmosphere

The first unambiguous evidence of a neutral atmosphere on Io was provided by Voyager 1/IRIS, which detected a signature of sulfur dioxide vapor (SO$_2$). IRIS covers a range from 180 cm^{-1} to 1900 cm^{-1}, with spectral resolution of 4.3 cm^{-1} and spatial resolution as good as 50 km. Figure 7.1 shows the SO$_2$ discovery spectrum from 1250 to 1450 cm^{-1} (Pearl et al. 1979). The feature between 1320 and 1380 cm^{-1} is identified as the v_3 band of SO$_2$. The strengths of other fundamental infrared active bands, v_1 (1152 cm^{-1}) and v_2 (518 cm^{-1}), are each a factor of 8 lower than the v_3 band. In fact, IRIS spectra show only a hint of the v_1 band, while v_2 is entirely masked by numerous unresolved features.

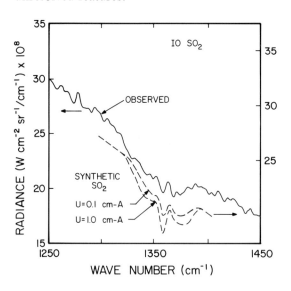

Fig. 7.1. Voyager/IRIS spectrum of Io (*solid line* and *left ordinate*) and its comparison with the synthetic SO$_2$ spectra (*broken lines* and *right ordinate*). The latter are for 0.1 cm-A and 1 cm-A of SO$_2$ gas at 130 K, and pressures 0.026 and 0.26 µbar, respectively. (Pearl et al. 1979)

Table 7.1. Io volcanic plume characteristics

Plume number	Volcano name	Detection by[a]	Location	Height[b] [km]	Width[b] [km]	Temperature[c] [K]
1	Pele	V1	19°S, 257°W	298	1242	654
2	Loki	V1, V2	17° – 19°N 301° – 305°W	148 – 338	350 – 445	245 – 450
3	Prometheus	V1, V2	3°S, 153°W	77	272	
4	Volund	V1	21°N, 177°W	98	94	
5	Amirani	V1, V2	27°N, 119°W	53 – 114	184 – 200	395
6	Maui	V1, V2	19°N, 122°W	68 – 105	343	200
7	Marduk	V1, V2	28°S, 210°W	64	160	
8	Masubi	V1, V2	45°S, 53°W	64	177	
9	Surt	V2	45°N, 338°W	?	50	600

[a] V1 and V2 represent Voyager 1 and Voyager 2 Imaging Subsystem observations (Smith et al. 1979a and 1979b; Strom and Schneider 1982).
[b] Dimensions are for clear filter observations, with the exception of Volund, whose dimensions are measured with a violet filter.
[c] Temperature data are from Voyager infrared spectrometer (Pearl and Sinton 1982).

The data shown in Fig. 7.1 represent average of seven spectra which cover a volcanic plume, Loki, a warm region at 290 K which occupied 10% of the IRIS field of view, and the remaining background at 130 K. The observed SO_2 could originate from any or all of these sources. Spectra taken over other volcanic plumes showed no evidence of SO_2. Table 7.1 lists characteristics of all volcanic plumes detected by Voyagers 1 and 2; some are hotter than Loki (such as Pele), and others colder (such as Maui). Thus, IRIS observations alone cannot distinguish between the sources of SO_2 as being either transient flow from active volcanoes, or a bound SO_2 atmosphere in equilibrium with SO_2 frost. Arguments can be made in favor of either source, as discussed below.

The infrared observations of SO_2, ionospheric data, and the Io plasma torus densities can all be interpreted satisfactorily in terms of a bound atmosphere of SO_2 controlled by its saturation vapor pressure. By comparing the SO_2 observations with synthetic spectra containing a single homogeneous layer of SO_2 at 130 K, Pearl et al. (1979) find that the best fit is obtained with an SO_2 abundance of 0.2 cm-atm with a probable error of a factor of 2 (Fig. 7.1 indicates the bounds of SO_2 abundance to be between 0.1 and 1.0 cm-atm). Surface pressure and density corresponding to the 0.2 cm-atm SO_2 are, respectively, 10^{-7} bar and 5×10^{12} cm^{-3}. The above values of the surface pressure and density are surprisingly close to the saturation vapor pressure and density of SO_2 in equilibrium with SO_2 frost/ice at 130 K, i.e., 1.4×10^{-7} bar and 10^{13} cm^{-3}, respectively (Wagman 1979; Fig. 7.2). This lends support to the presence of a bound atmosphere of SO_2 on Io. In other words, the data of Fig. 7.2 are weighted heavily toward the major contribution to the SO_2 signal being from the background (which was at 130 K), rather than the other two sources

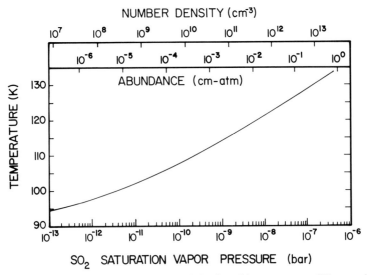

Fig. 7.2. SO_2 saturation vapor pressure behavior with temperature, (Wagman 1979). The corresponding surface number density (cm^{-3}) and column abundance (cm – atm) are indicated on the *upper abscissae*. (After Kumar and Hunten 1982)

(volcanic plume and the hot region). Indeed, the lack of detection of SO_2 signature over other plumes can be explained by the fact that these observations were done near the terminator or at night where the temperatures are in the 110 – 95 K range. The corresponding saturation vapor densities of SO_2, which lie in the $10^{10} - 10^7$ cm^{-3} range (vapor pressure range of 10^{-10} to 10^{-13} bar), are too small to be detected by Voyager/IRIS. On the other hand, it can also be argued that the SO_2 signature in Fig. 7.1 was caused mainly by transient flow from volcanic plume Loki, and that it could not have been detected over other plumes as they covered only a small fraction of the IRIS field of view. Thus, presence of a bound atmosphere of SO_2 on Io cannot be established unambiguously on the basis of the IRIS data in the Loki region alone. Additional evidence in favor of an SO_2 atmosphere is provided indirectly by the identification of the 4.1 μm feature in the ground-based observations of Io as SO_2 frost (Cruikshank 1980; Pollack et al. 1978). This would surely imply presence of SO_2 vapor in equilibrium with the frost at the appropriate surface temperature. Furthermore, the dayside ionospheric structure on Io can be interpreted satisfactorily if one were to invoke the presence of an SO_2 atmosphere with surface pressure of 10^{-9} bar, which is identical to the saturation vapor pressure of SO_2 at the temperature of the occultation point, 110 K. Except for some details, the nightside electron density profile can also be interpreted with an SO_2 atmosphere. Discussion of the ionosphere is presented later. The most persuasive evidence for a bound atmosphere on Io comes from the fact that to maintain the observed densities of sulfur and oxygen ions in the Io plasma torus, an SO_2 surface density exceeding 10^{11} cm^{-3} is needed (Kumar 1982). This is equivalent to the SO_2 saturation vapor density at 110 K, which can be taken to represent a reasonable global average temperature for Io. The atmosphere with an SO_2 surface density of 10^{11} cm^{-3} is generally referred to as the "thick" atmosphere.

The absence of any SO_2 signatures in IUE spectra, absence of polar caps, and possible subsurface cold trapping of SO_2 have led Lane et al. (1979), Butterworth (1980), Fanale et al. (1980) and Matson and Nash (1983) to argue in favor of a "thin" atmosphere with surface densities on the order of $10^8 - 10^9$ cm^{-3}, or of a lack of global SO_2 atmosphere on Io. Belton (1982), however, has demonstrated on the basis of a band absorption model for SO_2 longward of 2280 Å, that the IUE data were interpreted incorrectly with a continuum absorption model; hence the absence of an SO_2 signature can be treated as the "lower" limit to the SO_2 abundance on Io. The polar caps could be eroded by energetic charged particle impact as Io is immersed in Jupiter's intense magnetospheric environment. Migration of the polar cap material to the warmer latitudes is also a possibility, but is not fully understood. Particle and meteoritic impact would also expose any SO_2 frost/ice buried at shallow subsurface levels.

Although it is not fully demonstrated, the weight of the arguments seems to lie in favor of a global, "thick" atmosphere of SO_2 on Io, with its surface pressure determined by the saturation vapor pressure of SO_2 in equilibrium with SO_2 frost.

Volcanic Origin of SO_2

Io is constantly heated by strong tidal forces exerted by Jupiter, resulting in strong volcanic activity (Peale et al. 1979). This tidal interaction is related to the fact that the orbital period of Europa is twice that of Io. This 2 to 1 resonance causes Europa to force some eccentricity into Io's orbit. As a consequence, Jupiter exerts a tidal oscillation on Io resulting in its heating, and subsequently releasing the energy in the form of volcanic eruptions. The total power output from Io's surface is between 10^{13} and 10^{14} W. The tidal heating mechanism for volcanic activity on Io is starkly different from that for terrestrial volcanoes. The major heat source for the latter lies in the accretional energy, and decay of short halflife (Al^{26}) and long halflife (U^{238}, Th^{239}, etc.) radioactive elements. The intense volcanic activity on Io causes its surface to remain perpetually juvenile. This is evident from the fact that its surface is devoid of any impact craters. The absence of craters implies a resurfacing rate of 10^{-3} to 10 cm yr^{-1} (Johnson and Soderblom 1982) by the debris from volcanoes in the form of dust, particulates, and snow. This translates into many thousand metric tons s^{-1} of volcanic material ejected from Io. By way of comparison, one of the most intense terrestrial volcanoes, El Chichon, put out a total of 13.5 megatons of volcanic material (sulfur compounds and ash) in its most active period, 28 March – 4 April 1982; or roughly 20 tons s^{-1}, on the average (Evans and Kerr 1982). The SO_2 input into the atmosphere by the March, 1984, Mauna Loa eruption was roughly 200,000 tons (A. Krueger 1984, personal communication). From the decline in the ultraviolet signals of SO_2, measured by Pioneer Venus Orbiter from 1978 through 1983, Esposito (1984) has surmised the possibility of episodic volcanism on Venus. He estimates that the total aerosol injection into the middle atmosphere of Venus during this period was 200 megatons. It is thus apparent that Io possesses the most intense volcanism in the solar system.

The enormous resurfacing rate due to volcanism also implies that most of the light volatiles must have been driven out and lost from Io over geologic time. In fact, its surface density of 3.5 gm cm^{-3} makes it appear more rock-like (such as the Moon) than other Galilean satellites such as Ganymede and Callisto which, with a density of 2 gm cm^{-3}, are presumably half rock-half ice. Sulfur, which is one of the most abundant elements in the solar system ($S/H = 1.88 \times 10^{-5}$), is forced out of Io in some form and not lost directly. According to a model proposed by Smith et al. (1979a, b), the upper crust consists of sulfur (both from lava flows and as pyroclastic deposits) and SO_2. Tidal heating brings liquid sulfur dioxide in contact with molten sulfur at shallow subsurface levels. There the liquid SO_2 begins to boil and expand up one or several ducts toward the cooler surface, eventually reaching its triple point (198 K, 10 mb). Further expansion of SO_2 vapor results in a plume of SO_2 snow. The fate is expected to be same even if one were to begin with SO_2 vapor from the interior of Io. The geyserlike plume is expected to be nearly symmetrical, resembling an umbrella. The eruption velocities are on the order of 1 km s^{-1}, but still well below the escape velocity of Io (2.56 km s^{-1}). Such scenarios apply to several of the plumes detected by Voyager, such as Prometheus. Direct sulfur vapor volcanoes are also possible, as might be the case with extremely hot plumes of Pele. Thus the SO_2 frost/snow on Io appears to result from direct SO_2 snow from volcanic plumes, condensation of volcanic SO_2 vapor on contact with the cooler surrounding surface, and extrusion of liquid SO_2 through surface fractures. Figures 7.3 and 7.4 show examples of volcanic features, surface characteristics and evidence for sulfur dioxide, and molten sulfur flows on Io.

7.2.2 Minor Constituents

With the exception of the ν_3 band of SO_2, the Voyager infrared spectrum is devoid of any definite signatures. This is consistent with the earlier conclusion that Io's atmosphere is volatile-poor or -deficient. The upper limits of the various volatiles, deduced from their lack of detection in the Voyager infrared spectra, are given in Table 7.2. The H_2S upper limit is obtained from an analysis of the light curves of β Scorpii occultation by Io in 1971. The occultation occurred near the terminator, where the temperature is ~ 110 K. Assuming the

Fig. 7.3. The volcanic plume of Pele rising above the limb of Io. Another plume seen in this image is the bright spot toward the center. The image was taken through a clear filter from Voyager and was digitally enhanced by Morabito et al. (1979; Courtesy NASA/Voyager Project)

H_2S refractivity to be 6.3×10^{-4}, upper limits to the H_2S surface pressure and abundance are found to be 2×10^{-8} bar and 0.07 cm-atm, respectively (Pearl et al. 1979).

7.2.3 Neutral Cloud and Plasma Torus

In the extended environment of Io, emissions of neutral sodium, potassium, and oxygen have been detected by ground-based spectrometers, while hydrogen has been detected at the Lyman-α wavelength (1216 Å) both by Pioneer 10 as well as Voyager ultraviolet instruments. In addition, numerous magneto-

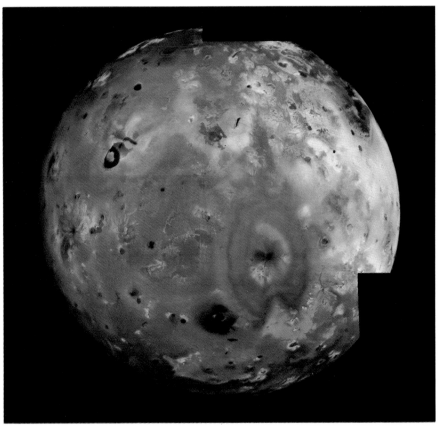

Fig. 7.4. Ejecta from the volcanic eruption of Pele can be seen as the large heart-shaped feature to the lower right of center in this mosaic of Io's surface. Many other volcanic features can likewise be spotted over the entire surface. The fluid-like appearance of the surface is presumably indicative of lava flows. The true color of Io is perhaps various pale shades of greenish yellow, with some areas possibly yellow-orange (Young 1984). Resolution is 8 km/line pair. (Courtesy NASA/Voyager Project)

spheric ion emissions have been observed by various instruments on Voyager. A summary of all the species detected so far, along with their concentrations in the extended region around Io, is provided in Table 7.3. Those details of the observations which are relevant to the aeronomy of Io are discussed below.

Intense emissions of sulfur and oxygen ions were detected by Voyager UVS, mainly at 685 Å and 833 Å, from a toroidal ring in the orbit of Io around Jupiter (Fig. 7.5). Additional signatures of sulfur and oxygen ions were observed at longer wavelengths with IUE. A synopsis of Voyager and IUE observations of the plasma torus is given in Table 7.4, while a summary of ground-based spectrometer data is presented in Table 7.5. The Voyager data demonstrate convincingly that the torus emissions are centered on the mag-

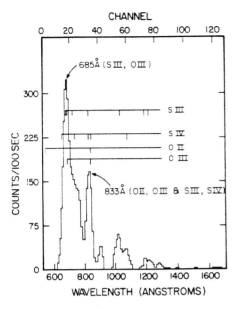

Fig. 7.5. Intense emissions of sulfur and oxygen ions from the Io plasma torus, detected by Voyager/UVS. The effects of instrumental scattering and the sky background have been removed. (After Boradfoot et al. 1979)

Table 7.2. Atmospheric composition at the surface of Io[a]

Constituent	Band/wavelength of detection [cm^{-1}]	Mixing ratio[b]
Major		
Sulfur dioxide, SO_2	v_3 1320 – 1380 v_1 1152 v_2 518 (ambiguous)	≥ 0.99
Minor (upper limits only)		
Carbonyl sulfide, COS	v_1 859	1.9×10^{-3}
Carbon disulfide, CS_2	v_3 1535	1.8×10^{-4}
Sulfur trioxide, SO_3	v_2 497	3.3×10^{-4}
Hydrogen sulfide, H_2S[c]		3.5×10^{-1}
Carbon dioxide, CO_2[d]	v_2 667	7.5×10^{-4}
Ozone, O_3	v_1 1042	9.5×10^{-3}
Nitrous oxide, N_2O[c]	v_1 589	3.7×10^{-2}
Water, H_2O[d]	Rot., 254	4.6×10^{-2}
Methane, CH_4[c]	v_4 1306	5.0×10^{-2}
Ammonia, NH_3[c]	v_2 931	7.0×10^{-3}
Hydrogen chloride, HCl	Rot., 206	1.5×10^{-2}

[a] After Pearl et al. (1979).

[b] Volume mixing ratios are calculated by assuming that the major atmospheric gas is SO_2, its abundance at the surface is 0.2 cm atm in a region where the temperature is 130 K; the corresponding surface pressure and density are 10^{-7} bar and 5×10^{12} cm^{-3}, respectively.

[c] Limits are improvements over values of Fink et al. (1976). H_2S limit is based on stellar occultation, assuming $T_{gas} = 100$ K.

[d] Improvements over values of Bartholdi and Owen (1972).

Table 7.3. Composition of Io's environment

Constituent	Concentration [cm^{-3}]	Experimental technique[a]	Reference
Ionosphere			
electrons, e$^-$ (dawn, 5.5 h)	9×10^3	P 10/RSS	1, 2
electrons, e$^-$ (dusk, 17.5 h)	6×10^4		
Plasma torus			
Singly ionized sulfur, S II	44	V/UVS	3
	260	V/PWS	4
Doubly ionized sulfur, S III	160	V/UVS	5
	260	V/PWS	4, 5
Triply ionized sulfur, S IV	216	V/UVS	5
	100	V/PWS	4, 5
Singly ionized oxygen, O II	49	V/UVS	5
	80	V/PWS	4, 5
Doubly ionized oxygen, O III	336	V/UVS	5
	100	V/PWS	4, 5
Triply ionized oxygen, O IV	<17		
S III or O II; S II or O III; and S_2^+ or SO_2^+	Total 2000	V/PWS	6
Total electrons, e$^-$	2000 – 4000	V/PRA	7
Neutral cloud			
Sodium, Na	10	Ground-based 5890 and 5896 Å	8
Potassium, K		Ground-based 7665 and 7699 Å	9, 10
Oxygen, O		Ground-based 6300 Å	11
Hydrogen, H		P 10/UVP (1216 Å)	12
		V/UVS (1216 Å)	13

[a] P 10/RSS: Pioneer/10 Radio Science (S-Band); V/UVS: Voyager/Ultraviolet Spectrometer; PWS: Plasma Wave; PRA: Planetary Radio Astronomy; P 10/UVP: Pioneer 10 Ultraviolet Photometer.
[1] Kliore et al. (1974); [2] Kliore et al. (1975); [3] Broadfoot et al. (1979); [4] Bagenal and Sullivan (1981); [5] Shemansky and Smith (1981); [6] Bridge et al. (1979); [7] Warwick et al. (1979); [8] Brown (1974); [9] Trafton (1975); [10] Trafton (1977); [11] Brown (1981); [12] Judge and Carlson (1974); [13] Broadfoot et al. (1981b).

netic equatorial plane of Jupiter. The cross-sectional diameter of the plasma torus is found to be 2 R_J, which is consistent with the fact that the plasma torus has an excursion of 1 R_J above and below the orbital plane of Io due to Jupiter's magnetic dipole offset of 10°. Figure 7.6 illustrates a model of Io torus along with a model fit to the intense 685 Å emission observed by Voyager/UVS. The source of sulfur and oxygen ions presumably lies in the SO_2 atmosphere of Io. Since the abundance of all volatiles except SO_2 in Io's atmo-

Table 7.4. Spacecraft and satellite observations of the Io plasma torus

Experiment	Feature	Constituent	Remarks
V/UVS[a]	685 Å	S III (680 Å, 683 Å and 700 Å) and O III (703 Å)	Minor contribution from S IV (657 Å), S III (724 Å and 729 Å) and S IV (745 Å)
V/UVS[a]	834 Å	O II (834 Å) and O III (835 Å)	Minor contribution from S III (835 Å) and S IV (816 Å)
V/UVS[a]	1018 Å, 1198 Å	S III	Not as prominent as the ones at 685 Å and 834 Å
V/UVS[a]	1070 Å	S IV	Contribution from S III (1077 Å)
IUE[b]	1198 Å	S III	
	1256 Å	S II	
	1406 Å	S IV	Marginal detection
	1664 Å	O III	Marginal detection

[a] Voyager/Ultraviolet Spectrometer Observations (Broadfoot et al. 1979).
[b] International Ultraviolet Spectrometer Observations (Moos and Clarke 1981).

Table 7.5. Ground-based observations of the Io plasma torus

Feature	Constituent	Reference
6716 Å	S II	1
6731 Å	S II	1
3722 Å	S III	2
6312 Å	S III	3
9531 Å	S III	4
3726 and 3729 Å	O II	5

[1] Kupo et al. (1976); [2] Morgan and Pilcher (1981); [3] Brown (1981); [4] Trauger et al. (1979); [5] Pilcher and Morgan (1979).

sphere is quite low (Table 7.2), only SO_2 is expected to undergo photolysis and particle impact dissociation, resulting in the production of sulfur and oxygen, amongst other things.

7.2.4 SO_2 Photochemistry and Model Atmosphere

For the thick atmosphere of SO_2 discussed earlier, there is sufficient column abundance of SO_2 available on the dayside of Io for photolysis to occur at or near the surface. SO_2 absorbs solar photons below 4000 Å in a complex system

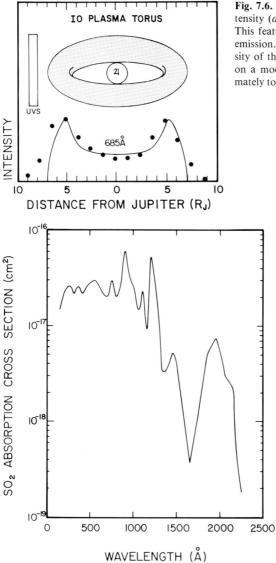

Fig. 7.6. Variation of the 685 Å emission intensity (*dots*) with the Jovicentric distance. This feature consists, primarily, of the S III emission. Also shown is the predicted intensity of the 685 Å emission (*solid line*) based on a model torus which is shown approximately to scale. (After Broadfoot et al. 1979)

Fig. 7.7. SO$_2$ photoabsorption cross-sections as a function of wavelength

of bands with an underlying continuum. The absorption cross-sections for the wavelength region most relevant to Io's SO$_2$ photochemistry, 2500 Å, are plotted in Fig. 7.7. Photolysis of SO$_2$ produces S, O, SO, O$_2$, and SO$_2^+$. A complete list of relevant reactions is presented in Table 7.6. Significant chemical pathways of the SO$_2$ photochemistry are shown schematically in Fig. 7.8. Much of the photochemical discussion here is based on Kumar's (1982) comprehensive study.

Table 7.6. SO$_2$ Photochemistry in Io's atmosphere[a]

Reactions	Rates[b]	Reaction number	Reference
SO$_2$ + $h\nu$ → SO + O	5.6×10^{-6}	R 1	1, 2
SO$_2$ + $h\nu$ → S + O$_2$	2.9×10^{-6}	R 2	1, 2
SO + $h\nu$ → S + O	6.0×10^{-6}	R 3	3
O$_2$ + $h\nu$ → O + O	9.5×10^{-8}	R 4	2
SO$_2$ + $h\nu$ → SO$_2^+$ + e$^-$	4.8×10^{-8}	R 5	4
S + $h\nu$ → S$^+$ + e$^-$	5.8×10^{-8}	R 6	5
O + $h\nu$ → O$^+$ + e$^-$	1.1×10^{-8}	R 7	6
SO + $h\nu$ → SO$^+$ + e$^-$	2.3×10^{-8}	R 8	7
O$_2$ + $h\nu$ → O$_2^+$ + e$^-$	2.3×10^{-8}	R 9	6
SO$_2$ + O + M → SO$_3$ + M	$3.4 \times 10^{-32} \exp(-1120/T)$	R 10	8
SO + O + M → SO$_2$ + M	3.0×10^{-33}	R 11	8
S + S + M → S$_2$ + M	2.0×10^{-33}	R 12	9
O$_2$ + O + M → O$_3$ + M	$1.4 \times 10^{-33} (T/300)^{-2.5}$	R 13	10
O + O + M → O$_2$ + M	$3 \times 10^{-33} (T/300)^{-2.9}$	R 14	10
S + O$_2$ → SO + O	$2.0 \times 10^{-11} \exp(-2820/T)$	R 15	11
SO + SO → S + SO$_2$	$5.8 \times 10^{-12} \exp(-1760/T)$	R 16	8
O$_2$ + SO → SO$_2$ + O	$7.5 \times 10^{-13} \exp(-3250/T)$	R 17	8
SO$_2^+$ + e$^-$ → SO + O	1×10^{-7}	R 18	12

[a] After Kumar (1982).
[b] Rate constants in cm^3 s^{-1} 2-body reactions, in cm^6 s^{-1} for 3-body reactions; photolysis and photoionization rates [s^{-1}] correspond to zero optical depth.
[1] Golomb et al. (1962); [2] Thompson et al. (1963); [3] Phillips (1981); [4] Wu and Judge (1981); [5] McGuire (1968); [6] Henry and McElroy (1968); [7] no data available, rates assumed same as for O$_2$ from Henry and McElroy (1968); [8] Baulch and Drysdale (1976); [9] Fair and Thrush (1967); [10] Liu and Donahue (1975); [11] Von Homann et al. (1968); [12] rate assumed.

The absorption by SO$_2$ of solar photons with wavelengths shortward of 2250 Å results in the production of SO and O, while that below 2070 Å yields S and O$_2$. Below 1000 Å, SO$_2$ is ionized to produce SO$_2^+$, i.e.,

$$SO_2 + h\nu\,(\lambda < 2250 \text{ Å}) \rightarrow SO + O \tag{R1}$$

$$+ h\nu\,(\lambda < 2070 \text{ Å}) \rightarrow S + O_2 \tag{R2}$$

$$+ h\nu\,(\lambda < 1000 \text{ Å}) \rightarrow SO_2^+ + e^- , \tag{R5}$$

(Reaction numbers correspond to those in Table 7.6)

Fluorescence yields are measured to be 1 at 2250 Å dropping to 0.1 at 2150 Å (Hui and Rice 1972); quantum yield for reaction R2 is ≤ 0.5 at 1849 Å (Driscoll and Warneck 1968).

S and O$_2$ react rapidly to form SO and O, this is the major pathway for the loss of S and O$_2$:

$$S + O_2 \rightarrow SO + O . \tag{R15}$$

SO formed in the above reaction and R1 reacts with itself and O$_2$ to recycle SO$_2$ and form S and O. Removal of SO by photodissociation occurs below

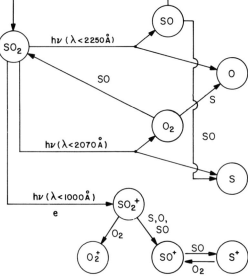

Fig. 7.8. Photochemistry of SO_2 on Io

2400 Å, but is not as important as the self-reaction. SO removal reactions are summarized below.

$$SO + SO \rightarrow SO_2 + S \tag{R16}$$

$$SO + O_2 \rightarrow SO_2 + O . \tag{R17}$$

Numerous 3-body reactions (R10 – R14) are important near the surface where the pressure is relatively high. The absorption of solar ultraviolet radiation by SO_2 with its subsequent dissociation is the principal mechanism for the upper atmospheric heating on Io. The thermal structure is calculated in the following manner.

The appropriate one-dimensional heat conduction equation to use, is (after Kumar 1980):

$$\frac{\partial}{\partial z}\left(\kappa \frac{dT}{dz}\right) + H_v(z) + H_c(z) - R(T, z) = 0 , \tag{7.1}$$

where $\kappa = 0.494\, T^{1.329}$ erg cm^{-1} K^{-1} is the coefficient of thermal conductivity for SO_2; T is the temperature in K; $H_v(z)$ and $H_c(z)$ are, respectively, heating rates at altitude z due to solar EUV and charged particles; and $R(T, z)$ is the radiative cooling rate for SO_2. $R(T, z)$ can be expressed as

$$R(T, z) = n^2(z) k_v\, h\nu \exp(-h\nu/kT) , \tag{7.2}$$

where $n(z)$ is the atmospheric density at altitude z; $h\nu$ is the energy of radiated photon; and k_v is the rate coefficient for vibrational quenching of SO_2. Since

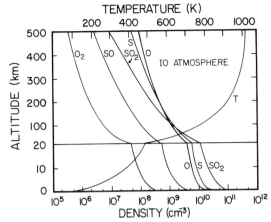

Fig. 7.9. Model atmosphere of Io resulting from SO_2 photochemical reactions shown in previous figure. The lower O concentrations correspond to a chemically active surface. (Based on Kumar 1980 and 1982)

no data for k_v of SO_2 are available, it is assumed to be the same as for CO_2 in the context of Mars (Chamberlain and McElroy 1966), i.e.,

$$k_v = 1.9 \times 10^{-9} \exp(-82.8\, T^{-1/3}).$$

(7.3)

The SO_2 atmosphere cools in the SO_2 infrared bands v_1, v_2, and v_3 whose strengths are in the ratio $1:1:8$.

The solutions of the above expressions for thermal structure, and the continuity equations yield the thermal profile and species distributions as shown in Fig. 7.9. The exospheric temperature is calculated to be approximately 1000 K. The model assumes a solar zenith angle of 81°, and a surface temperature of 110 K to simulate the geometry of Pioneer 10 radio occultation. The corresponding saturation vapor density of SO_2 at the surface is 1.2×10^{11} cm^{-3}, so that unit optical depth in SO_2 occurs close to the surface. The bulk of photolysis goes on within 10 km above the surface. At higher altitudes SO_2 is recycled faster (R16 and R17) than its loss due to photolysis. Diffusive separation of species occurs above $\sim$125 km. Since S and O have, respectively, two and four times the scale height of SO_2, they become the dominant species above $\sim$125 km.

The atmospheric model shown in Fig. 7.9 assumes that the fluxes of SO, O_2, S, and O are zero at the surface, i.e., they are simply products of SO_2 photochemistry (SO_2 flux is 2×10^{11} cm^{-2} s^{-1} at the surface). Once produced, the products are lost through photochemical reactions, or they escape the atmosphere.

The model just discussed does not account for the possibility that the surface may be continually renewed by lava flows, pyroclastic or meteoritic deposits, upwelling, and particle erosion. If the oxidation rate is faster than the surface renewal rate, the atmospheric oxygen atoms could form O_2 by a 3-body recombination reaction at the surface (Kumar 1982; Kumar and Hunten 1982). An atmospheric model for this scenario of chemically active surface developed by Kumar (1982) gives $O_2/SO_2 = 0.15$, or $O_2 = 10^{10}\,\text{cm}^{-3}$ at the surface. These O_2 densities are large enough to explain the nightside ionosphere, as will be discussed in the following section. Sulfur would be a minor constituent in this model, since S combines rapidly with O_2 (R15). Atomic oxygen continues to be the dominant constituent above the homopause.

7.2.5 Ionospheric Measurements

The only measurements available of the ionosphere of Io are those done by Pioneer 10 in 1973, using the radio occultation technique (Fig. 7.10; Kliore et al. 1974 and 1975). The "Entry" occultation measurements were done at dusk (17.5 h local time; downstream) and gave a peak electron concentration of $6 \times 10^4\,\text{cm}^{-3}$ at an altitude of roughly 100 km. The "Exit" measurements were done near the dawn terminator (5.5 h local time; upstream) and resulted in a peak electron concentration of $9 \times 10^3\,\text{cm}^{-3}$ at an altitude of 50 km. The peak altitudes are based on an "average" radius of Io of 1875 km determined in the above experiment (Kliore et al. 1975). However, since the correct radius is 1815 ± 5 km (Morrison 1982), the heights of maxima in the electron concentrations of the entry and exit occultation data are uncertain by at least 60 km.

The earlier theoretical speculations on the ionosphere of Io were made before simultaneous information concerning the neutral atmosphere was available. These models were based on a neutral atmosphere composed of NH_3 (McElroy and Yung 1975; Gross and Ramanathan 1976; Johnson et al. 1976), Na (McElroy and Yung 1975), and Ne (Whitten et al. 1975). Voyager observations have shown that the only gas with relatively large concentration in the atmosphere of Io is SO_2. An ionosphere composed of Ne^+, even if it were

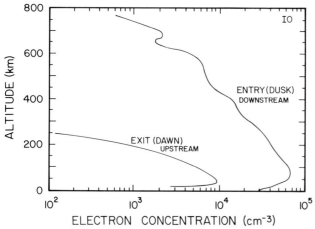

Fig. 7.10. Pioneer 10 measurements of the Io electron density profiles at dusk (local time 17.5 h) and dawn (local time 5.5 h) (Based on data from Kliore et al. 1974 and 1975)

produced in large quantities, is not probable, as Ne^+ reacts rapidly with SO_2 to form a short-lived molecular ion. Na^+ could be a potentially important candidate for the nightside, as discussed later. Since the volcanic activity on Io detected by Voyager in 1979 could be episodic, it cannot be said with certainty whether the atmosphere of Io 6 years earlier, at the time of Pioneer 10 observations, was also composed of SO_2. However, the persistence of volcanism between Voyager 1 and Voyager 2 epoch, and the time-independent nature of the mechanism for such volcanism would tend to argue in favor of a continuous volcanic activity on Io. Although the intensity of such volcanism might wax and wane, the concentration of SO_2 in the atmosphere is limited, in any event, by the saturation vapor pressure of SO_2 in equilibrium with SO_2 frost on the surface.

7.2.6 Ionospheric Models

The absorption of solar photons below 1000 Å by SO_2 leads to the production of SO_2^+ ions (R5). The loss of these ions occurs either by electron recombination (R18), or through the following sequence of charge exchange reactions (also shown in Fig. 7.8):

$$SO_2^+ + O \rightarrow SO^+ + O_2 \tag{R19}$$

$$SO_2^+ + S \rightarrow SO^+ + SO \tag{R20}$$

$$SO_2^+ + SO \rightarrow SO^+ + SO_2 \tag{R21}$$

$$SO_2^+ + O_2 \rightarrow O_2^+ + SO_2 . \tag{R22}$$

Removal of SO_2^+ by charge exchange with O (R19) may constitute the most effective loss mechanism for SO_2^+, since O is a major gas at ionospheric heights in the model with chemically inactive surface (Fig. 7.9). SO^+ produced by R19, R20, and R21 undergoes the following reactions

$$SO^+ + SO \rightarrow S^+ + SO_2, \tag{R23}$$

followed by

$$S^+ + O_2 \rightarrow SO^+ + O . \tag{R24}$$

Since the rate constants for reactions R19 – R24 are not known accurately under Ionian conditions, it is not apparent whether SO_2^+, SO^+, or S^+ would be the major ion. However, calculations fit the data best with SO_2^+ as the major ion at the peak, (as discussed later). At higher altitudes, the identity of the ions might indeed be different.

Using the formulation developed in the Ionosphere chapter and a model atmosphere for Io (Fig. 7.9), it is found that the solar EUV alone produces a peak electron concentration of $10^4 \, cm^{-3}$ at conditions appropriate to the Pioneer 10 entry occultation. Since this value is a factor of 6 too small compared to the data, corpuscular ionization is presumed responsible for Io's iono-

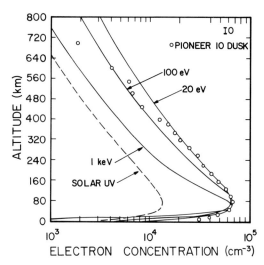

Fig. 7.11. Comparison of the Pioneer 10 dusk measurements of the Io iono-sphere with several models assuming charged particle and solar EUV ioniza-tion in an SO_2 atmosphere. (Kumar 1980)

sphere. The Io plasma torus is found to have an ion concentration of $2000\ cm^{-3}$ with mean ion energy of $30-40\ eV$ and the mean electron energy of $10\ eV$ with corresponding electron flux of $4 \times 10^{11}\ cm^{-2}\ s^{-1}$ (Bridge et al. 1979; Bagenal and Sullivan 1981). Using the above information for the ioniz-ing flux, a model atmosphere as shown in Fig. 7.9, and the formulation devel-oped in Chapter 6 on Ionosphere, one can calculate the electron concentration profile. Kumar (1980) found that the best fit to the Entry occultation data is obtained with $20\ eV$ electrons with a flux of $1.2 \times 10^{10}\ cm^{-2}\ s^{-1}$, and an SO_2 density at the surface of $1.2 \times 10^{11}\ cm^{-3}$ (Fig. 7.11). The electron peak is calculated to be at around the $100\ km$ level, using the formulation developed in Chapter 6. The peak electron concentration is given by the following equation (see Sect. 7.4.5 on Titan Ionosphere for electron profile formula-tion).

$$[N_e]_{max} = 0.6 \left(\frac{F(\infty)E}{35\ \alpha H} \right)^{1/2}, \tag{7.4}$$

which yields $[N_e]_{max} \approx 6.2 \times 10^4\ cm^{-3}$ with $F(\infty) = 1.2 \times 10^{10}\ cm^{-2}\ s^{-1}$, $E = 20\ eV$, $\alpha \approx 10^{-7}\ cm^3\ s^{-1}$ (R18), and $H = 61.5\ km$ at $100\ km$, assuming $R_{Io} = 1875\ km$ so as to be compatible with the Pioneer 10 height scale. [For a complete $N_e(z)$ profile calculation, electron cross-sections may be taken from Orient and Srivastava (1984), where $Q_{SO_2^+}(20\ eV) = 1.6 \times 10^{-16}\ cm^2$.] It is in-conceivable that S^+ could be the major ion at the peak, since Eq. (7.4.) with $\alpha_{S^+} \approx 10^{-12}\ cm^3\ s^{-1}$, gives $[N_e]_{max} \approx 10^7\ cm^{-3}$. There is only a slight likeli-hood of obtaining the correct peak electron concentration even with SO^+ as the ion at the peak. The needed electron flux in the above model is compatible with the Io plasma torus data, and the needed SO_2 density is equivalent to the SO_2 saturation vapor pressure in equilibrium with SO_2-ice at $110\ K$. The surface temperature in the Entry occultation region is indeed 110 K. Earlier models of Johnson et al. (1976) also invoked particle ionization, and assumed a "thin" atmosphere with surface densities a factor of 100 lower, i.e., $10^9 \, cm^{-3}$. The model, however, used a mean electron energy of 100 eV. Besides, the atmospheric scale heights were not known at the time of the Pioneer observations.

Interpretation of the exit occultation observations is in a less satisfactory state because of the lack of supporting data on neutral atmosphere and relevant plasma conditions. The exit measurements refer to the morning terminator, where the surface temperature is ≤ 95 K, with corresponding SO_2 saturation densities $\leq 10^8 \, cm^{-3}$ (Fig. 7.2). With these densities, ordinary ionization mechanisms fail to produce the observed electron concentration of $10^4 \, cm^{-3}$ at the peak. On the other hand, whereas the shadow height during the exit occultation is at 25 km, the peak in the electron concentration is at around 50 km (or higher, because of the uncertainty in the height scale). Thus, the solar EUV ionization could still be considered as a viable source. The SO_2 density needed to produce an electron concentration of $10^4 \, cm^{-3}$ at the peak is, however, $5 \times 10^{10} \, cm^{-3}$, i.e., more than 100 times greater than the SO_2 vapor pressure at the local surface temperature. Larger SO_2 densities could be provided by transport across the terminator, since the SO_2 surface densities on the dayside are $10^{11} - 10^{12} \, cm^{-3}$. The topside scale height of the observed ionosphere is 25 km, which is about half the value expected for a solar EUV-generated ionosphere. Perhaps ram pressure caused by Jupiter's co-rotating magnetosphere plasma is responsible for compressing the ionosphere (Kumar and Hunten 1982).

Other possible candidates for ionization are Na^+ and O_2^+. Both could be abundant in Io's atmosphere. The difficulty with any Na^+ produced, either by solar EUV or charged particles, is that these ions would quickly react with the neutral atmosphere, resulting in short-lived molecules or cluster ions with typical lifetimes of less than 100 s at the peak. As discussed earlier, O_2 could be produced in large quantities, with surface densities $\approx 10^{10} \, cm^{-3}$ in a model with chemically active surface. Since the saturation vapor pressure of O_2 is 1 bar at 90 K, there is no danger of its condensation on the nightside of Io. Thus O_2^+ ions remain potentially important for interpreting the exit ionosphere profile.

There is also a distinct possibility that interaction of the magnetospheric plasma with Io's atmosphere might play an important role in determining the characteristics of the ionosphere measured by the exit occultation. This is because those observations were done on the upstream side. Although SO_2 densities might be small, appropriate combination of charged particle energy and flux could still produce the observed characteristics. Simultaneous observations of the neutral atmosphere and the ionosphere can remove most of the current ambiguities in the interpretation of Io's ionospheric data. Such observations are planned from the Galileo Orbiter, using the occultation techniques.

7.3 Europa, Ganymede, Callisto, and Saturn's Rings

A ground-based observation of stellar occultation by Ganymede indicated that the surface pressure on Ganymede might be at least 10^{-6} bar (Carlson et al. 1973). A highly sophisticated monitoring of an ultraviolet stellar occultation from Voyager, however, placed a rigid upper limit of 10^{-11} bar on Ganymede's surface pressure (Broadfoot et al. 1979). Although no atmospheric measurements of Europa and Callisto are available, there is a distinct possibility of a thin atmosphere on at least some of the Galilean moons, as explained in the following paragraphs.

Europa, Ganymede, and Callisto are all covered with varying degrees of water-ice. Europa's surface is nearly 100% water-ice-covered. Moreover, its surface is virtually devoid of any impact craters; it is, however, criss-crossed with linear features which look like cracks or fractures. It seems that Europa's surface is continually renewed either by resurfacing or melting of water-ice. Ganymede, on the other hand, is 20 to 65% covered with water-ice; its surface is substantially cratered and fractured. The surface of Callisto is 5 to 25% covered with water-ice. It is heavily cratered and ringed, much like the Earth's moon. Some of the above-mentioned features of these Galilean moons are clearly evident in Fig. 7.12. The composition of the surface is, by weight, $\geq 95\%$, 90%, and 30 to 90% H_2O-ice at, respectively, Europa, Ganymede, and Callisto. The subsolar (full disc brightness) temperatures are approximately 124, 135, and 150 K, respectively, for Europa, Ganymede, and Callisto. The mean temperatures, on the other hand, are all in the range of 95 to 105 K. The saturation vapor pressure of water vapor in equilibrium with ice at the above temperatures ranges from 3×10^{-18} mb (at 95 K) to 5×10^{-8} mb (at 150 K), with an intermediate value of 6×10^{-10} mb at 135 K. From the above characteristics, it seems that there is some potential for generating an atmosphere on these moons, since water is photolyzed easily at the ultraviolet wavelengths. The solar EUV dissociation of H_2O eventually leads to the formation of O_2 in the following sequence of reactions.

$$H_2O + h\nu \,(\lambda < 2000 \text{ Å}) \rightarrow OH + H$$

followed by

$$OH + OH \rightarrow H_2O + O$$

and

$$O + OH \rightarrow O_2 + H \,.$$

Oxygen is recycled by

$$O_2 + h\nu \,(\lambda < 2600 \text{ Å}) \rightarrow O(^1D) + O(^3P)$$

and,

$$O_2 + h\nu \,(\lambda < 2600 \text{ Å}) \rightarrow O(^3P) + O(^3P) \,.$$

Fig. 7.12. A family portrait of the four Galilean moons of Jupiter, clockwise from the upper left: Io, Europa, Ganymede, and Callisto. The relative sizes, colors, and reflectivities are approximately correct. This mosaic was prepared from images of the moons taken by Voyager from a distance of $\sim 10^6$ km from each moon. (Courtesy NASA/Voyager Project)

The solar photons, with wavelengths shorter than 1530 Å and 2020 Å, respectively, produce oxygen atoms with energy ≥ 0.63 eV in the last two reactions. These atoms possess enough energy to overcome Ganymede's gravitational field (escape velocity, $v_E = 2.75$ km s^{-1}) and would eventually escape. If the production exceeds the loss, O_2 would build up in the atmosphere. Yung and McElroy (1977) calculate that an "average" partial pressure of 2×10^{-9} mb of H_2O would be needed to satisfy the above condition, and would result in a microbar surface pressure of O_2 on Ganymede. This would occur over a relatively short period of approximately 10,000 years. Once the O_2 optical depth is large enough to prevent additional photodissociation of H_2O (both O_2 and H_2O absorb in the same wavelength range), O_2 pressure would be stabilized. The biggest difficulty with this scenario is that the needed average partial pressure of H_2O (2×10^{-9} mb) would require a "mean" surface temperature of approximately 140 K. The actual mean surface temperature of Ganymede, ~ 100 K, would result in a dramatically smaller value of the water vapor pressure of $\sim 10^{-17}$ mb. Even at the subsolar temperature of 135 K, the water vapor pressure would be $\sim 6 \times 10^{-10}$ mb. Even allowing for the uncertainties due to the extrapolation in the saturation vapor pressure and the albedo of the moon is not expected to alter these results substantially. Thus, it is evident that formation of an O_2 atmosphere on Ganymede in excess of the Voyager/UVS upper limit is virtually impossible. A slight likelihood, however, does exist of accumulating an O_2 atmosphere with pressure in the 0.01 to 1-µb range on the surface of Callisto. As discussed in the context of the Jovian CO (Chap. 5 on Photochemistry), erosion of water-ice of the moons by the magnetospheric charged particles (Lanzerotti et al. 1978; Johnson et al. 1981) is another mechanism for producing a thin atmosphere around the Galilean moons. A microbar level oxygen atmosphere would give rise to ionospheric electron concentrations of the order of $(1 - 10) \times 10^3$ cm^{-3}, as can be concluded from the earlier discussion on Io's ionosphere.

Finally, impact of charged particles, meteoritic bombardment, as well as neutralization of the ionospheric H^+, can all result in a tenuous atmosphere of hydrogen in the plane of Saturn's rings (see Chap. 6.2.6, Ionosphere).

7.4 Titan

Titan, the second largest moon in the solar system, was discovered by Christian Huygens in 1655. Nearly a quarter millenium later in 1908, Comas Solas detected limb darkening on Titan, thus providing the first indication of the presence of an atmosphere on a planetary satellite. Modern scientific measurements of Titan's atmosphere began with Kuiper's (1944) detection of CH_4 in the visible (6190 Å) and near infrared photographic spectra. Kuiper (1952) deduced a CH_4 abundance of 0.2 km-am, without, however, accounting for curve-of-growth or temperature effects. Later, Trafton (1972) detected CH_4 in the 3 ν_3 band (1.1 µm) and also re-analyzed Kuiper's 6190 Å data. Taking into account self-broadening (i.e., pure CH_4 atmosphere), Trafton deduced a surface pressure of 16 mb, or CH_4 abundance of 1.6 km-am. Lutz et al. (1976) analyzed the weak absorption bands in the visible (where pressure dependence is insignificant), and concluded that the surface pressure is 200 mb, and that CH_4 is a minor constituent on Titan. Any of the numerous candidates with large cosmic abundances, such as Ne, Ar, N_2, could be the background gas. All the above models, however, incorrectly assumed a simple scattering layer without including aerosols above it. Interest in Titan was re-kindled following radio measurements at 3 mm (Conklin et al. 1977), which indicated surface temperature in excess of 200 K. Considering pressure-induced opacity in an assumed N_2 atmosphere (with a small proportion of H_2), Hunten (1978) concluded that a surface pressure of 20 bar would be needed to maintain a 200 K surface temperature. Soon afterward, Jaffe et al. (1980) found that the 3 mm results were erroneous; they in turn determined a surface temperature of 87.9 K from 6 cm VLA (Very Large Array) measurements. Hunten's (1978) model atmosphere gives a surface pressure of 1 bar for $T = 87$ K. Nitrogen normally eludes detection in the infrared since it has no permanent dipole moment. Being a homonuclear diatom it does not have a rotational-vibrational spectrum due to dipole transitions. Even in the ultraviolet, nitrogen is difficult to detect because of the low intensity of its various bands. Furthermore, observations from the Earth suffer from contamination by the Earth's own nitrogen and hydrogen Lyman-α. The detection of N_2 on Titan had to await the Voyager 1 flyby in 1980.

In addition to CH_4, pre-Voyager ground-based infrared spectroscopy observations revealed the presence of C_2H_6 at 12.2 µm (Gillett et al. 1973; Danielson et al. 1973), CH_3D at 8.7 µm and C_2H_4 at 10.5 µm (Gillett 1975), and perhaps C_2H_2 at 13.7 µm and C_3H_8 at 13.4 µm (Gillett 1975).

From the above-mentioned observations, it had become abundantly clear that Titan is a unique celestial body whose atmosphere had some possibility of resembling our own in a primitive stage.

7.4.1 Voyager Overview

On 12 November, 1980, Voyager 1 made a close encounter with Titan, coming to within 3900 km of its surface. The surface was, however, concealed well below the omnipresent cover of orange brown clouds and haze layers. An

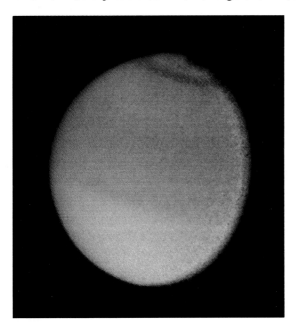

Fig. 7.13. An image of Titan taken by Voyager 1 wide-angle camera. Outstanding features are high altitude haze, detached polar hood, and a north-south asymmetry in the albedo. (Courtesy NASA/Voyager Project)

Table 7.7. Atmospheric properties of Titan

Property	Value	Reference
Surface		1
Pressure	1496 ± 20 mb	
Temperature	94 ± 0.7 K	
Effective temperature	86 K	
Measured lapse rate		1
0 – 3.5 km	1.38 ± 0.1 K km^{-1}	
3.5 km	0.9 ± 0.1 K	
42 km	0	
Tropopause		1
Height	42 km	
Pressure	127 mb	
Temperature	71.4 ± 0.5 K	
Stratopause		1
Height	200 km	
Pressure	0.70 mb	
Temperature	170 ± 15 K	
Homopause		2
Height	925 ± 70 km	
Density	2.7×10^{10} cm^{-3}	
Eddy diffusion coefficient	$1.0^{+2.0}_{-0.7} \times 10^8$ cm^2 s^{-1}	
Exosphere		2
Temperature		
evening terminator	176 ± 20 K	
morning terminator	196 ± 20 K	
Critical level	1600 km	
Density at critical level	9.3×10^6 cm^{-3}	3
Incident solar flux	1.1×10^{-2} of flux at 1 AU	

[1] Lindal et al. (1983); [2] Smith et al. (1982); [3] Bertaux and Kockarts (1983).

asymmetry in the north-south albedo was detected, such that the region south the equator was lighter brown/orange than that north of the equator. A detached polar hood was seen, beginning at around 65°N latitude and entirely surrounding the moon. Many of these features can be noticed in the wide-angle image taken by the Voyager cameras (Fig. 7.13). In the upper atmosphere, a distinct haze layer was detected in the ultraviolet spectra at around 700 – 1000 km. The atmosphere was found to be predominantly N_2, with trace amounts of hydrocarbons and other organics. The pressure at the surface is 50% greater than at the Earth's surface, and the temperature closer to the VLA data. A summary of the atmospheric properties as determined by the various Voyager instruments is presented in Table 7.7, and also included in Fig. 7.14. The following sections discuss in detail the thermal structure, atmospheric composition, aeronomy – including neutral and charged particle chemistry, and atmospheric evolution of Titan.

7.4.2 Thermal Structure

The currently accepted thermal profile of Titan is shown in Fig. 7.14 and the temperatures at various levels are listed in Table 7.7. The refraction of radio waves at 13 cm and 3.6 cm by the atmosphere of Titan has been analyzed to yield an atmospheric temperature of 94 ± 0.7 K at the surface, assuming a pure N_2 atmosphere. The lapse rate up to 3.5 km is found to be 1.38 ± 0.1 K km^{-1}, which is entirely consistent with a 100% N_2 atmosphere as discussed below.

The dry adiabatic lapse rate for an ideal gas is given by

$$\Gamma_d = g/C_p \tag{7.5}$$

where,

$g = 1.354$ m s^{-2} at Titan's surface, and

$C_p = 1069$ J kg^{-1} K^{-1}, the specific heat of N_2

Thus,

$\Gamma_d = 1.27$ K km^{-1} at Titan's surface.

However, N_2 is close to phase transition on Titan; it does not behave like an ideal gas. Taking into consideration the van der Waals corrections due to intermolecular forces and molecular compressibility (Handbook of Chemistry and Physics), Lindal et al. (1983) derive a lapse rate of 1.40 K km^{-1} at Titan's surface, which is in agreement with the measured value of 1.38 K km^{-1}.

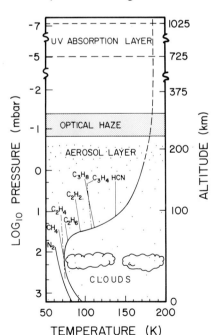

Fig. 7.14. Titan thermal structure. Interpolation between the lower atmospheric measurements by the infrared and radio techniques and the upper atmospheric measurements by UV occultation is shown by a *broken line*. Also shown are the locations of clouds and hazes deduced from the Voyager data. Condensation curves of HCN etc. are for their mixing ratios in Titan's atmosphere, e.g., HCN with 2×10^{-7} mole fraction will condence at 15 mb level. (Based on the Voyager IRIS and RSS data; and Maguire et al. 1981)

The radio measurements, however, yield only the scale height information. Thus, they provide only $T/\bar{m}$, i.e., the ratio of temperature (T) to the mean molecular weight ($\bar{m}$). Knowledge of either T or $\bar{m}$ is essential to infer atmospheric properties from the data. In principle, the radio data can be normalized against the infrared data, since infrared opacities can be used to determine the temperature at some reference level. In practice, however, infrared measurements at Titan were capable of yielding only a wide range of possible temperatures for the following reasons. Most of the infrared opacity at Titan results from CH_4, whose abundance has not been determined directly. Any indirect measure of the CH_4 abundance applies to the region above the clouds, and models assume uniform vertical mixing of CH_4 in the stratosphere. Using a wavelength region $(200 - 540 \text{ cm}^{-1})$ which does not critically depend on the nature of the opacity sources, Samuelson et al. (1981) have determined a lower limit of 94 ± 0.4 K for the air temperature immediately above the surface (the surface in this model is 0.5 K warmer). The upper limit is highly uncertain, with a maximum value of 98 K, and a nominal acceptable value of 95 K (Samuelson et al. 1981; Hunten et al. 1984). Therefore, despite large uncertainties, the infrared data can be satisfactorily normalized to the radio measurements, yielding atmospheric temperature of 94 K at the surface. This implies a mean molecular weight of ~28, i.e., the atmosphere is composed of essentially pure N_2. Small quantities of CH_4 and Ar at the surface can be tolerated within the limits of these measurements.

By modeling the dependence on temperature of the shape of the 1304 cm^{-1} band, and the emission levels at 1304 cm^{-1} and 1260 cm^{-1}, Samuelson et al. (1981) have obtained a temperature profile in the $0.1 - 5$ mb range. Both infrared and the radio data yield temperature of 170 ± 15 K at the stratopause which is located around the 1-mb level.

Like Jupiter and Saturn, Titan has an "information gap" region in the middle atmosphere (200 km to 1250 km or 0.1 mb to 5 nb) where no temperature measurements are available. At a radial distance of 3840 km and beyond, the Voyager ultraviolet spectrometer measured the column abundances of N_2, CH_4, and C_2H_2 in a solar occultation experiment (Broadfoot et al. 1981a). The column abundances can be used to provide scale heights, and subsequently temperatures (see Chap. 2 on Thermal Structure). The angular diameter of the Sun projected into the atmosphere of Titan was 3 km during Voyager 1 entry occultation, and 17 km during exit, both values being considerably less than the atmospheric scale height (65 km) at these altitudes. The observed transmission characteristics are consistent with the principal absorbers being N_2 for $\lambda < 800$ Å, and CH_4 and C_2H_2 for $\lambda > 800$ Å (Smith et al. 1982). The measurements yielded the following values for the exospheric temperature in the equatorial region (3°N): 196 K near the morning terminator, and 176 K near the evening terminator. Thus a globally averaged value for the exospheric temperature on Titan is 186 ± 20 K.

The N_2 density at the upper end of the "information gap" region, ($R = 3840$ km), is found from the solar occultation experiment to be

$2.7 \times 10^8 \, \text{cm}^{-3}$. In order to reconcile it with the atmospheric density near the lower end, $(R = 2750 \, \text{km}, 4.5 \times 10^{16} \, \text{cm}^{-3})$, it is imperative that the mean temperature between 2750 km and 3840 km radii be 165 K. Since the stratopause $(R = 2775 \, \text{km})$ temperature is 170 K, and exosphere $(R = 3840 \, \text{km})$ temperature is 186 K, a temperature of 165 K in this height range would imply considerable structure in the thermal profile. This is not unreasonable considering the presence of aerosols of all sizes over an extensive height range on Titan. A working model of the atmospheric number density can then be constructed in the following manner.

For 2575 km $< R <$ 2750 km, use the measured n and T (Lindal et al. 1983).
For 2575 km $< R <$ 3840 km, use the hydrostatic law of densities with
  $n(2750 \, \text{km}) = 4.5 \times 10^{16} \, \text{cm}^{-3}$, $T(2750 \, \text{km}) = 165 \, \text{K}$ and constant up to 3840 km.
For $R >$ 3840 km, use again the hydrostatic law, but with $n(3840 \, \text{km})$
  $= 2.7 \times 10^8 \, \text{cm}^{-3}$, as derived in the previous step, and $T = 186 \, \text{K}$, the average exospheric temperature.

The hydrostatic law of densities is given below

$$n(R_2) = n(R_1) \left(\frac{T_1}{T_2} \right) \exp \left[- \int_{R_1}^{R_2} dR/H(R) \right] \tag{7.6}$$

where

$$H(R) = \frac{kT(R)}{\bar{m}g(R)} \quad \text{(scale height)} . \tag{7.7}$$

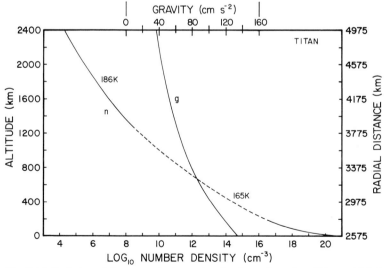

Fig. 7.15. Variation of the atmospheric number density and gravity of Titan with height. *Broken line* interpolation has the same meaning as in the previous figure

With $T=$ const between 2750 km and 3840 km, $\bar{m}=28$ AMU for an essentially N_2 atmosphere, the scale height simply becomes a function of gravity, which varies with R as,

$$g(R) = g(R_1)(R_1/R)^2 \ . \tag{7.8}$$

With the above assumptions, Eq. (7.6) becomes

$$n(R_2) = n(R_1) \exp\left[-\left(\frac{R_1}{H(R_1)} - \frac{R_2}{H(R_2)}\right)\right] \ . \tag{7.9}$$

The same expression holds for $R > 3840$ km, except $T = 186$ K in Eq. (7.7) for $H(R)$. The distribution of individual constituents in the exosphere (e.g., H or H_2) can also be calculated using Eq. (7.9), but with $H(R)$ to mean the scale height of the atom/molecule in question.

The resulting model atmosphere is shown in Fig. 7.15. It should be regarded only as an engineering model until thermal structure measurements in the middle atmosphere are available, e.g., from an entry probe such as the one under consideration for the Cassini Project sometime in the future.

Global Variation

For the most part, infrared measurements of the spatial distribution of temperature indicate no significant latitudinal or diurnal variation at any heights in the atmosphere (Flasar et al. 1981). The results at three different pressure levels are summarized in Table 7.8. The relative global uniformity in temperature in the troposphere is reasonable since the radiative time constants up to 50 km are large compared to the 30-year seasonal cycle, and the slow spin rate, as discussed below.

The radiative time constant at the surface, τ_r, can be approximated by the ratio of the atmospheric heat capacity to the globally averaged solar energy input (Gierasch et al. 1970; Smith et al. 1981; Leovy and Pollack 1973), i.e.,

$$\tau_r = \frac{(p_0 T_0)/g/C_p}{(\phi_s/4)(1-A)} \ , \tag{7.10}$$

where p_0 and T_0 are mean pressure and temperature at the surface (1500 mb and 94 K); ϕ_s is the solar constant at Titan's orbit (1.55×10^4 erg cm^{-2} s^{-1}); other quantities are defined elsewhere.

The above expression yields $\tau_r \approx 4 \times 10^9$ s which is approximately four times the length of the season (10^9 s) on Titan. The calculations for τ_r in the stratosphere are complicated due to the presence of aerosols and the uncertainties in the abundances of the hydrocarbons. Nevertheless, Flasar et al. (1981) have calculated that τ_r decreases to a value around 3×10^7 s at the stratopause level (~ 1 mb). Thus, larger global variations in the temperature are expected, and are likewise inferred from the data at these altitudes (Table 7.8). No lon-

Table 7.8. Latitudinal and diurnal variations of temperature in Titan's atmosphere

Wave number used in analysis [cm^{-1}]	Corresponding height region	Corresponding pressure region	Temperature variation	
			Latitudinal	Diurnal
1304	130 km	1 mb	Equator to pole: ~20 K Midlatitude: Southern hemisphere ~4 K warmer	
200	40 km (tropopause)	100 mb	None definite. Suggestion of slight increase from South to North	None discernable
530	Surface to lower troposphere	1500 – 50 mb	About equator: symmetric, Poles: ~3 K cooler	35°N: 1.2 K. 0 to 40°: Daytime slightly warmer in the North. None in the radio data (−8.5° and +6.2°)

gitudinal variations are seen, as is implied by the fact that τ_r at all altitudes is greater than the diurnal cycle of Titan, which is 16 days.

7.4.3 Atmospheric Composition

The presently known constituents and their abundances in Titan's atmosphere are listed in Table 7.9. Discussion about noteworthy species is presented below.

Nitrogen

Although N_2 had been proposed as a candidate responsible for the pressure broadening of the observed CH_4 emission bands, its direct detection was made by the Voyager Ultraviolet Spectrometer. Figure 7.16 shows the various N_2, NI, and NII emissions detected in the 600 – 1700 Å range. The intensities of the principal emission features are listed in Table 7.10. In addition, as discussed earlier, the solar occultation data below 800 Å can be interpreted

Table 7.9. Atmospheric composition of Titan

Constituent	Mole fraction	Band or wavelength of detection	Reference
Major			
Molecular nitrogen, N_2	0.98 – 0.82	Radio and IR deduction from the $\bar{\mu}$ considerations; UV: RYD, BH, LBH	1, 2
Argon, Ar^{36}	$<6 \times 10^{-2}$ (at 3900 km)	1048 Å	3
Methane, CH_4	$\sim 2 \times 10^{-2}$ (at 100 mb)	1304 cm^{-1}	4
	$(8 \pm 3) \times 10^{-2}$ (at 3700 km)	<1200 Å	5
Minor			
Nitrogen (atomic), NI	?	1134 Å multiplet	2
Hydrogen (atomic), HI	$<10\%$ (at 3900 km)	1216 Å	2
Hydrogen (molecular), H_2	$(2 \pm 1) \times 10^{-3}$	360, 600 cm^{-1}	6
Neon, NeI	$<1 \times 10^{-2}$	736 Å	3
$C-H$ *Group*			
Ethane, C_2H_6	2×10^{-5}	822 cm^{-1}	4
Propane, C_3H_8	4×10^{-6}	748 cm^{-1}	7
Acetylene, C_2H_2	2×10^{-6} (at 1 mb) $10^{-3} - 2 \times 10^{-2}$ (at 3300 km)	≥ 1400 Å	4, 5
Ethylene, C_2H_4	4×10^{-7}	950 cm^{-1}	8
Methylacetylene, CH_3C_2H	3×10^{-8}	325, 633 cm^{-1}	9
Diacetylene, C_4H_2	$10^{-8} - 10^{-7}$	220, 628 cm^{-1}	8
$C-N$ *Group*			
Cyanogen, C_2N_2	$10^{-8} - 10^{-7}$	233 cm^{-1}	8
$C-H-N$ *Group*			
Hydrogen cyanide, HCN	2×10^{-7}	712 cm^{-1}	8
Cyanoacetylene, HC_3N	$10^{-8} - 10^{-7}$	500, 663 cm^{-1}	8
$C-O$ *Group*			
Carbon monoxide, CO	6×10^{-5}	~ 6350 cm^{-1} (3 – 0) Rot. Vib.,	10
Carbon dioxide, CO_2	$1.5^{+1.5}_{-0.8} \times 10^{-9}$	667 cm^{-1} v_2 Q-branch	11

[1] Lindal et al. (1983); [2] Broadfoot et al. (1981a); [3] Strobel and Shemansky (1982); [4] Hanel et al. (1981a); [5] Smith et al. (1982); [6] Samuelson et al. (1981); [7] W. Maguire 1981, pers. commun.; [8] Kunde et al. (1981); [9] Maguire et al. (1981); [10] Lutz et al. (1983); [11] Samuelson et al. (1983).

properly only with N_2 as the major absorbing gas. The UV results, however, refer to the upper atmosphere ($P < 1$ µb); they cannot predict the bulk composition of the troposphere. For that, one has to rely on the indirect evidence provided by the combined radio occultation refractivity and infrared opacity

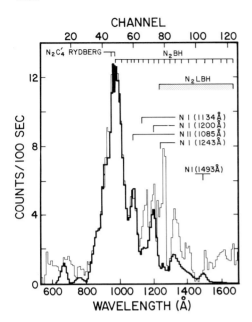

Fig. 7.16. The principal N, N^+ and N_2 emissions from Titan detected by Voyager/UVS. The effects of instumental scattering and Lyman-α have been removed from the spectra. A synthetic model based on the electron impact ionization of N_2 is shown by *heavy lines*. (Strobel and Shemansky 1982)

Table 7.10. Significant UV emissions in Titan's spectra[a]

Species	Observed intensity (R)		Predicted intensity (R)	
	Disc average	Bright limb	$T_e = 10^6$ K	$T_e = 2.0 \times 10^5$ K[b]
N_2RYD(0,0) 958 Å	12	21	212	110
N_2RYD(0,1) 981 Å	8.6	21	36	18.7
N_2RYD(0,2) 1003.2 Å	7.0	5.8	4.2	2.2
N_2BH$(1, v'')$	25	97	33	33
N_2LBH system	96	290	180	426
NI (1493 Å)	15	49	61	21
NI (1243 Å), (marginal)	8.0	27	39	11.6
NI (1200 Å), (marginal)	30	102	166	43
NI (1134 Å)	7.9	27	24	12
NII (1085 Å)	12	42	88	15

Upper limits to the volume mixing ratios at $R = 3900$ km[c]

Ne (736 Å)	0.01
Ar (1048 Å)	0.06
H_2 (Lyman and Werner bands)	0.06
H (1216 Å)	0.10
CO (1447 Å: 4th positive)	0.05

[a] Adapted from Strobel and Shemansky (1982).
[b] With average dayside power input of 2×10^{10} W, and electron impact on optically thin N_2 gas. Electron temperature, $T_e = 2 \times 10^5$ K is equivalent to an auroral secondary electron flux varying as $E^{-1.4}$. $T_e = 10^6$ K is assumed appropriate for the primary excitation above the critical level (Strobel and Shemansky 1982).
[c] $[N_2] = 2 \times 10^8$ cm^{-3} at 3900 km. Upper limits are obtained from the spectral signatures at the indicated wavelengths.

analysis discussed previously. The tropospheric mean molecular weight derived from these data is consistent with an almost pure N_2 atmosphere. Small amounts of methane and argon are allowed by the uncertainty in the temperature determination.

Argon

No definite detection of argon in the atmosphere of Titan is available. A feature seen at 1044 ± 5 Å in a limb spectrum (Broadfoot et al. 1981a) is located suspiciously close to the 1048 Å resonance line of Ar^{36}. This feature was seen only at 5000 km radial distance. No signatures were detected for Ar lines between 866 and 876 Å. An upper limit derived by Broadfoot et al. is $Ar/N_2 < 6\%$ at 3900 km. In the homosphere, the ratio could be larger, since argon density would drop above the homopause (3500 km). Since argon has many strong emission lines in the ultraviolet, its nondetection, even in the presence of CH_4 and N_2 on Titan, casts doubts on the existence of this constituent in more than a few percent abundance.

The radio and thermal data discussed earlier can also yield some information on the argon abundance. The upper limit on the surface temperature is 95 K, which implies a mean molecular weight, $\bar{m} = 28.3$. Some gas heavier than N_2 is needed to account for this value of $\bar{m}$. Argon is an excellent candidate, being heavier than N_2 and in abundant supply in the solar system. $\bar{m} = 28.3$ implies $Ar^{36}/N_2 = 4\%$ in the troposphere. It should be emphasized that the interpretation of the radio occultation data does not require any argon in the atmosphere of Titan (Eshleman et al. 1983; Lindal et al. 1983; Eshleman 1982).

From the solar system abundances of elements, a ratio of 9.2% is predicted for Ar/N_2. This argon should be primordial Ar^{36}, not Ar^{40}. The mole fraction of the latter (Ar^{40}) − which is presumably derived from the decay of K^{40} in Titan's core − would be no more than 0.01% (Owen 1982).

Methane, Other Organics, and Clouds

The Voyager infrared spectra revealed the presence of numerous hydrocarbons and other organics in the $C - N - H$-group, as shown in Fig. 7.17. The CH_4 mole fraction at the tropopause is found to be 2%. This is also the region where clouds are seen all over Titan. No measurements of the CH_4 abundance below the clouds are available. Since the abundance of CH_4 and the formation of clouds is interrelated, the following discussion deals with both issues together.

The saturated vapor mixing ratio of CH_4 at the tropopause (71.4 K, 127 mb) is 3%. There is little probability of global CH_4 condensation, since the measured value of CH_4 there is only 2%. The clouds near the tropopause

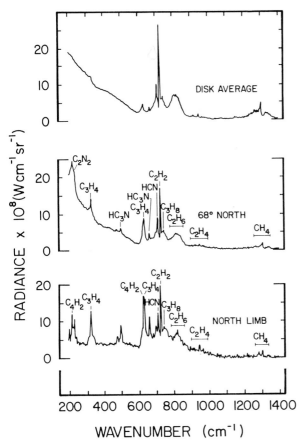

Fig. 7.17. Averaged infrared spectra of Titan for midlatitudes (disc average), high northern latitudes (68°N) and off the north polar limb situation. (After Kunde et al. 1981)

cannot be made up of CH_4. At the surface, a CH_4 mixing ratio of 12% would be needed for saturation (CH_4 triple point, 90.7 K). However, as discussed earlier, the observed lapse rate up to 4 km is consistent with a pure N_2 atmosphere. Even allowing for the lower limit to the surface temperature (94 ± 0.7 K) gives $\bar{m} = 27.8$, which would imply a maximum of 2% CH_4 at the surface. Again, there is no possibility of CH_4 condensation at the surface. In the polar regions, where the surface is 3 K cooler, there is some likelihood of the CH_4 condensation. Above 4 km, the lapse rate drops to 0.9 K km^{-1} and continues down to 0 at the tropopause. Such a transition at 4 km is reflective of an atmosphere which is in convective equilibrium below 4 km and radiative equilibrium above it (Eshleman et al. 1983). Again, to interpret the observed thermal structure above 4 km, saturated levels of methane are not required (Eshleman et al. 1983). In conclusion, CH_4 does not appear responsible for the global clouds on Titan, its mixing ratio at the tropopause is 2%, and it is not expected to be much greater deeper in the atmosphere. In the upper atmosphere, photochemical and diffusive processes control the mixing ratio of CH_4 and its products.

The clouds cannot be composed of the other hydrocarbons or N_2, as their partial pressures are lower than the saturation vapor pressures at these heights. Some possibility does exist of high altitude haze formation by the organics. It is important to recognize that most of the vapor pressure data currently available are for vapor over pure liquids or solids. In the presence of binary, tertiary, or other mixtures, the vapor pressures can be substantially altered, leading to different condensation scenarios than those discussed here.

CH_4 and its photolysis products C_2H_2, C_2H_6, and C_2H_4 are found to be globally uniform. HC_3N and C_2N_2, on the other hand, are detected only in the north polar region ($\geq 60°N$). HCN, CH_3C_2H, and C_4H_2 also decrease from the northern high latitudes to the southern high latitudes. Such latitudinal behavior appears to be related to the lifetimes and the production mechanisms of these species, as discussed later.

H_2 and H

Voyager instruments detected both H_2 and H in Titan's spectra. H_2 was observed as an infrared absorption feature in the troposphere, whereas H was inferred from an extensive torus of Lyman-α emission. The mean thermal velocities of H and H_2 at the temperature of Titan's critical level are, respectively, 1.8 and 1.3 km s^{-1} (additional details on critical level can be found in Chap. 2 on Thermal Structure). They are thus comparable to the planetary escape velocity of 2.6 km s^{-1} at the surface and 1.7 km s^{-1} at the critical level. The escape of these light gases in copious quantities from Titan's atmosphere is assured. The limiting flux,

$$\phi_1 = \frac{b_i f_i}{H_a} \tag{7.11}$$

(see Chap. 4 on Vertical Mixing) of H_2 at the homopause of Titan is calculated to be $(3.8 \pm 2) \times 10^9$ cm^{-2} s^{-1}. It represents the maximum allowable escape rate for H_2. It is comparable to the column integrated photolysis rate of CH_4 (next section). CH_4 is the most likely source of Titan's H and H_2. The escape flux of atomic hydrogen is expected to be similar to that of H_2.

CO and CO_2

Both of these gases could be either indigenous to Titan's primordial interior, or they could have formed as a result of the influx of some extra-satellite oxygen-bearing material. The first plausibility is discussed later in a possible scenario for the evolution of Titan's atmosphere. The chemistry subsequent to the introduction of extraplanetary H_2, O, or OH, has been presented in Chapter 5 on Photochemistry in the context of the Jovian carbon monoxide.

7.4.4 Chemistry

The chemical processes in the present-day Titan atmosphere are dominated by coupled reactions between molecules of H, C, and N. Some likelihood of oxygen chemistry indigenous to Titan also exists. The neutral species are produced through photochemical reactions, whereas ions are formed primarily as a result of charged particle precipitation, as discussed in the following subsection.

Methane Photochemistry

Because of the presence of copious amounts of N_2 as the background gas, the photochemistry of CH_4 in Titan's atmosphere differs somewhat from that in the Jovian atmosphere. Near and just below the homopause, CH_4 photolysis produces CH, $^1CH_2$ and $^3CH_2$, same as in the Jovian atmosphere (Chap. 5 on Photochemistry). However, on Titan, $^1CH_2$ is rapidly quenched by N_2 to $^3CH_2$, unlike on Jupiter, where $^1CH_2$ is the major photodecomposition product of CH_4. The sequence of reactions relevant to Titan is summarized below:

$$CH_4 + h\nu\,(\lambda \leq 1450 \text{ Å}) \rightarrow CH + H_2 + H \qquad\qquad (R\,25)$$

$$\rightarrow {}^3CH_2 + 2\,H \qquad\qquad (R\,26)$$

$$\rightarrow {}^1CH_2 + H_2 \qquad\qquad (R\,27)$$

followed by

$$^1CH_2 + N_2 \rightarrow {}^3CH_2 + N_2. \qquad\qquad (R\,28)$$

The column-integrated photodissociation rate of CH_4 at Titan is found to be $1.2 \times 10^9\,\text{cm}^{-2}\text{s}^{-1}$ due to direct photolysis and $4 \times 10^9\,\text{cm}^{-2}\text{s}^{-1}$ including catalytic photolysis and $R\,32 - R\,35$ (Yung et al. 1984).

The product CH subsequently undergoes the following reactions, leading to the formation of C_2H_4 and, eventually, C_2H_2.

$$CH + CH_4 \rightarrow C_2H_4 + H \qquad\qquad (R\,29)$$

followed by

$$C_2H_4 + h\nu\,(\lambda < 1700 \text{ Å}) \rightarrow C_2H_2 + H_2. \qquad\qquad (R\,30)$$

$^3CH_2$, on the other hand, reacts with itself to produce C_2H_2, i.e.,

$$^3CH_2 + {}^3CH_2 \rightarrow C_2H_2 + 2\,H. \qquad\qquad (R\,31)$$

$^3CH_2$ can also react with C_2H_2 to form methylacetylene, CH_3C_2H, or allene, $CH_2=C=CH_2$. C_2H_2 produced by reactions (R\,30) and (R\,31) flows down to the stratosphere where it can be photolyzed. Any surviving C_2H_2 condenses at the tropopause cold-trap from where it will eventually snow out of the atmosphere.

Photolysis of C_2H_2 in the stratosphere results in the formation of C_2H, which reacts preferentially with C_2H_2 to give diacetylene, C_4H_2, i.e.,

$$C_2H_2 + h\nu \,(2000 \text{ Å} \gtrsim \lambda \gtrsim 1450 \text{ Å}) \rightarrow C_2H + H \tag{R 32}$$

followed by

$$C_2H + C_2H_2 \rightarrow C_4H_2 + H. \tag{R 33}$$

C_4H_2 can further react with C_2H to produce polyacetylenes (Allen et al. 1980), i.e.,

$$C_4H_2 \xrightarrow{C_2H} C_6H_2 \xrightarrow{C_2H} C_8H_2 \xrightarrow{C_2H} \text{ etc.} \tag{R 34}$$

The polyacetylenes absorb sunlight up to 3200 Å, forming $C_{2n}H$ ($n = 1, 2, 3\ldots$). Subsequent photochemistry re-forms C_2H. At the appropriate concentration ratio of C_2H_2 and CH_4, the reaction of C_2H shifts in favor of CH_4, i.e.,

$$C_2H + CH_4 \rightarrow CH_3 + C_2H_2. \tag{R 35}$$

(Note that the above C_2H is produced not simply from C_2H_2, but also from polyacetylenes.) Also the activation energy of this reaction is 2 K cal mol^{-1}, so that it is not favorable at colder temperatures, such as those in the north polar region of Titan. There, reaction (R 33) with C_2H_2 would result in the formation of polyacetylenes.

The net result of reactions (R 32 – R 35) is catalytic decomposition of CH_4 to produce methyl radical, CH_3. Ethane is produced by a 3-body reaction, i.e.,

$$CH_3 + CH_3 + N_2 \rightarrow C_2H_6 + N_2 \tag{R 36}$$

as on Jupiter (where N_2 is replaced by H_2).

The C_2H_6 production rate is expected to be much greater than the C_2H_2 production rate because of greater photolysis rates above 1450 Å (which produce primarily C_2H_6) than below 1450 Å (which produce C_2H_2). Photolysis of C_2H_6 and subsequent reactions with CH_3 result in the formation of propane (C_3H_8), butane (C_4H_{10}), etc., as is the case in the Jovian atmosphere. The transport of C_2H_6 to the tropopause leads to its condensation, followed by rain-out. CH_4 is recycled mainly by reaction of CH_3 with H, as on Jupiter.

Based on the above CH_4 photolysis model, and using the available thermodynamic data on the $CH_4 - C_2H_6 - N_2$ system, Lunine et al. (1983) have estimated that Titan's surface should be covered with a 1-km-deep ocean of ethane. Since the vapor pressures of C_2H_2 are considerably lower at the corresponding temperatures, acetylene is expected to be primarily in the solid form, and could form in a layer 100 – 200 m thick at the bottom of the ethane liquid. For the previously mentioned limits on the atmospheric CH_4 abundance at the surface, Lunine et al. derive mole fractions of 13 to 25% for CH_4 and 5% for N_2 in the ethane ocean. This conclusion, however, should be regarded as tentative as the relevant thermodynamic data, such as vapor pressures, solubility etc., over a $CH_4 - C_2H_6 - N_2$ (and, perhaps even C_2H_2) ensemble are poorly known below 110 K. Any future probe to Titan (e.g., Cassini) must carry an adequate complement of instruments to determine Titan's surface characteristics.

Hydrocarbons and Nitrogen

N_2 is highly inert, and it does not react directly with the hydrocarbons in Titan's atmosphere. On the other hand, atomic nitrogen reacts readily with CH_4. N_2 is dissociated by the solar photons below 1000 Å, magnetospheric electrons, and galactic cosmic rays (Strobel and Shemansky 1982). The globally averaged flux of the resulting $N(^2D)$ and $N(^4S)$ atoms due to magnetospheric electron impact is $1.8 \times 10^9 \, cm^{-2} s^{-1}$, which is greater than that due to the solar UV and cosmic ray dissociation of N_2 by factors of 3 and 10, respectively. The chemical reactions following the production of nitrogen atoms are schematically shown in Fig. 7.18, and their highlights discussed below.

$N(^2D)$ reacts with CH_4 to yield $N(^4S)$, NH or HCN. The branching ratios of these reactions, which are unknown, determine the HCN production rate. The ground state $N(^4S)$ reacts with CH_3 and $^3CH_2$ (produced in the CH_4 photolysis) to yield HCN. HCN is photodissociated to yield CN, i.e.,

$$HCN + h\nu \rightarrow H(^2S) + CN(^2\pi) \, . \qquad (R37)$$

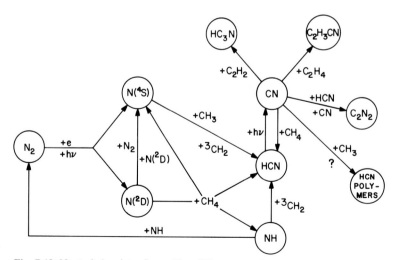

Fig. 7.18. Neutral chemistry for an $N_2 - CH_4$ system relevant to Titan

There is a great deal of controversy surrounding the bond dissociation energy of HCN. Davis and Okabe (1968) claim it is 5.2 eV (2384 Å), while Herzberg and Innes (1957) observed predissociation below 1780 Å. Since there are only fragmentary data on the absorption cross-section of HCN (Lee 1980; Herzberg and Innes 1957; Hilgendorf 1935), cross-sections for C_2H_2, which is isoelectronic with HCN, or HCl may be used for the wavelength region where HCN data are missing.

Reactions of CN with CH_4, CN or HCN, C_2H_2, and C_2H_4 yield, respectively HCN, cyanogen (C_2N_2), cyanoacetylene (HC_3N), and ethyl cyanide (C_2H_3CN). Because of the relatively large abundance of C_2H_2 in Titan's atmosphere, its reaction with CN produces relatively large quantities of HC_3N.

Aerosols

The reaction

$$CN + CH_4 \xrightarrow{k} CH_3 + HCN , \qquad\qquad\qquad (R38)$$

where $k = 10^{-11} \exp(-866/T) \ cm^3 \ s^{-1}$, Schacke et al. 1977,

has large activation energy of $2 \ k \ cal \ mol^{-1}$. Therefore, the above reaction is at least a factor of 2 less efficient at the colder temperatures of the north polar hood region than elsewhere on Titan. As a consequence, at those latitudes, formation of complex molecules, such as HC_3N, would be favored in reactions of CN with the ambient atmospheric gases. As discussed earlier, polyacetylenes would also be produced in large abundance in those regions. Indeed, both C_4H_2 and HC_3N were detected by Voyager/IRIS only in the north polar region (Fig. 7.16). In addition, polymers of HCN can form in the reaction between CN and CH_3 (Fig. 7.17). All these complex molecules, polyacetylenes, and HCN polymers give rise to photochemically or charged particle-generated smog in Titan's atmosphere. Sagan and Khare (1979) and Sagan et al. (1984) have observed the formation of reddish-brown powder, called "tholins" (mud), in experiments where "simulated" atmospheres of Titan and the Jovian planets were irradiated by disequilibrating energy sources such as ultraviolet light and charged particles. Because of physical limitations, these laboratory experiments cannot truly simulate planetary conditions. For example, the extremely low pressures where the energy is deposited on Titan cannot be used effectively in the laboratory simulation experiments. The actual energy deposited in Titan's atmosphere is generally much lower than that from laboratory irradiating sources. Furthermore, at Titan, the energy is available for billions of years, a condition the laboratory experiments cannot simulate. There is also the nagging problem of wall effects. In addition to the laboratory experiments, Cerceau et al. (1985) predict, from a study of the infrared spectra of various $H - C - N$ molecules, the presence of various nitriles in the atmosphere of Titan. Despite the obvious difficulties associated with the above-mentioned simulations, it seems virtually certain that Titan harbors conditions necessary for the formation of numerous highly complex aerosol molecules. The absorption of sunlight by the aerosols with subsequent formation of an inversion layer will result in strong static stability. This will in turn result in the formation of smog. Although the thermodynamic properties of the various above-mentioned aerosol candidates are not well understood, it is suspected that, due to the presence of impurities, their vapor pressures would be depressed far below those over pure parents. This would enhance the formation and retention of the aerosols in Titan's atmosphere. Finally, it is important to note that charged particle precipitation in the high latitudes may be even more effective than photochemical processes in producing organics and other complex hydrocarbons.

7.4.5 Ionosphere

Voyager 1 spacecraft was occulted by Titan on 12 November, 1980. An analysis of the differential dispersive frequency measurements done during the occultation revealed an upper limit of approximately 3000 electrons cm^{-3} on the evening terminator ($\chi = 94°$), and 5000 electrons cm^{-3} on the morning terminator ($\chi = 86°$), (Lindal et al. 1983). Considering the composition of Titan's neutral atmosphere, and the large magnetospheric electron energy input, it is an inescapable conclusion that Titan would possess a healthy ionosphere, especially when it lies inside the magnetosphere of Saturn. The best indirect evidence of an ionosphere on Titan comes from the Voyager plasma measurements, which indicated presence of a plasma wake region surrounding Titan in Saturn's rotating magnetosphere (Bridge et al. 1981). Since Titan does not possess an intrinsic magnetic field (Ness et al. 1981), it appears that the rotating magnetospheric plasma was deflected by currents induced in the ionosphere, and it formed a bipolar tail. Since the solar EUV does not produce large quantities of nitrogen ions on Titan, it is not an important ionization source. During Voyager 1 encounter, Titan was located within the magnetosphere of Saturn, although it can be encroached upon by the solar wind from time to time. The impact of magnetospheric electrons produces globally averaged values of 9×10^8 cm^{-2} s^{-1} of N$_2^+$, and 1.6×10^8 cm^{-2} s^{-1} of N$^+$ (Strobel and Shemansky 1982). These large ion production rates lead to the formation of numerous other ions by charge exchange, atom/molecule-ion interchange, and other processes. Laboratory measurements of most of the chemical reactions relevant to Titan's ionosphere have been carried out by Huntress et al. (1980). The ionospheric scheme is shown schematically in Fig. 7.19, and its salient features discussed below.

The N$_2^+$ ions react primarily with CH$_4$ producing CH$_2^+$, which are quickly neutralized. N$_2^+$ ions also react with H$_2$ to yield N$_2$H$^+$ ions, which in turn dissociate to N$_2$ and H. In effect, N$_2^+$ ions do not participate in the ion chemistry to any large degree. It is the N$^+$ ions which are the major driver of Titan's

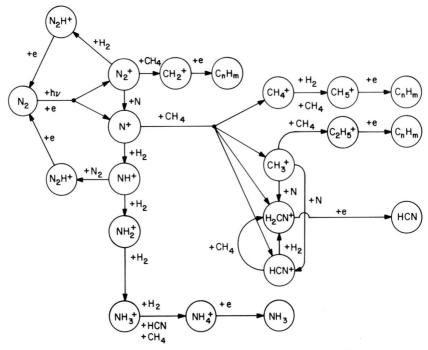

Fig. 7.19. Schematic representation of possible ionospheric chemistry on Titan

ionospheric chemical processes. The N^+ react with both H_2 and CH_4. The reaction with H_2 and subsequent reactions, however, produce only small concentrations of NH_n^+ ($n = 1, 2, 3, 4$) ions because of the low density of H_2 in Titan's atmosphere. The reaction with CH_4, on the other hand, yields CH_3^+, H_2CN^+, HCN^+, and CH_4^+ with branching ratios of roughly 1/2, 1/3, 1/10, and <1/20, respectively. Both CH_4^+ and CH_3^+ undergo further reactions with CH_4 and/or H_2 to yield heavier hydrocarbon ions, such as CH_5^+ and $C_2H_5^+$, as in the Jovian ionosphere. The CH_3^+ ions react also with atmospheric N to produce HCN^+ ions. The HCN^+ ions, in turn, are converted to H_2CN^+ ions on reaction with H_2 or CH_4. The terminal ion is therefore H_2CN^+, with HCN^+ not trailing too far behind in concentration. (At least one Voyager plasma measurement gives evidence of mass 28 ions in Titan's magnetic tail (Hartle et al. 1982). An excellent candidate would be H_2CN^+ or N_2^+ of ionospheric origin which is picked up by the rotating magnetosphere). The terminal ions are removed by electron recombination. In this manner, the ionosphere also becomes a source of HCN in the upper atmosphere.

The calculations for a model ionosphere of Titan are similar to those for Io, based on the charged particle ionization. The electron density profile is calculated by using the following expression, similar to that for the Io ionosphere

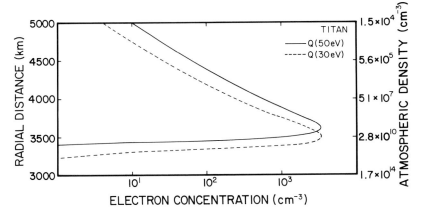

Fig. 7.20. Models of the ionosphere of Titan, assuming charged particle ionization with 50 and 30 eV electrons

$$N_e(z) = \left(\frac{n(z)F(\infty)EQ(E)}{35\,\alpha} \right)^{1/2} \exp\left(\frac{-n(z)HQ(E)}{2} \right).$$ (7.12)

For Titan, the values of the relevant parameters are: globally averaged magnetospheric energy input of 0.05 erg cm^{-2} s^{-1}, cross-section, $Q(E)$, for the production of N$^+$ on electron impact of N$_2$ of 5×10^{-18} cm^2 at 30 eV and 2.8×10^{-17} at 50 eV (Strobel and Shemansky 1982), electron recombination rate constant, α, for the terminal ion (H$_2$CN$^+$ or HCN$^+$) assumed to be the same as that for the hydrocarbon ions, CH$_5^+$ or C$_2$H$_5^+$, i.e., 3.9×10^{-6} cm^3 s^{-1} (Maier and Fessenden 1975), and a model atmosphere as discussed earlier in this chapter. The resulting ionospheric profiles are shown in Fig. 7.20. The peak concentration is found to be approximately 3500 electrons cm^{-3} at a radial distance of roughly 3500 to 3700 km (altitude 900 – 1100 km). The peak electron concentration is in agreement with the upper limit for either the evening or the morning terminator. In fact, the above model calculations are more germane to the morning terminator where the observed upper limit of electron concentration is 5000 cm^{-3}. This is due to the fact that the N$_2$ emissions were found to be more intense there than on the evening terminator, implying greater magnetospheric energy input on the morning side. The magnetospheric energy input assumed in the above calculations is the maximum value for the dayside, and it is expected to be somewhat less on the morning terminator, so that the peak electron concentrations would be somewhat lower than 3500 cm^{-3}. Other factors, such as the sweeping of plasma by the solar wind, etc. would also tend to lower the topside ion densities. Since the observed N$_2$ emissions are mainly confined to the sunlit hemisphere and that the excitation energy derives from the magnetospheric particle precipitation, the ionosphere is expected to be virtually nonexistent in the absence of this source. Therefore, on the nightside, and when Titan is outside the magnetosphere of Saturn, the ionosphere is expected to collapse.

7.4.6 Origin and Evolution of Titan's Atmosphere

The mass and escape velocity of Titan are in the same range as of the Galilean moons of Jupiter. Yet Titan alone is endowed with a substantial atmosphere. The reason for this may lie in the fact that, unlike the Galilean moons, Titan is only sometimes inside the planetary magnetospheric environment, which is, moreover, not quite as intense as the Jovian magnetosphere. Thus, stripping of the atmosphere by energetic charged particles, especially in the early phases of evolution, is not a serious threat. This implies that the accumulation of an atmosphere on Titan would have proceeded virtually unabated. Furthermore, Titan's location in the cooler part of the nebula from which it was accreted, appears to permit a more favorable proportion of rocks and ices for the formation of an atmosphere. In this part of the nebula, there is a greater likelihood of incorporating into Titan's interior ammonia and methane ices, in addition to the water-ice (Hunten et al. 1984). The latter (H_2O-ice) is most probably the only ice incorporated in the interior of the Galilean moons.

Three possible scenarios can be envisioned to explain the presence of a massive atmosphere of nitrogen on Titan: (i) direct capture of N_2 from the solar or proto-Saturnian nebula, (ii) devolatilization of nitrogen containing ices condensed from the nebula – the clathrate hydrate model, and (iii) photolysis of ammonia in the early phases of accretion.

Direct N_2 Capture

There is some controversy as to the form of nitrogen in the primordial nebula from which the planets and satellites were accreted. It is generally assumed that nitrogen was present in the form of NH_3. However, Lewis and Prinn (1980) have surmised on the basis of kinetic arguments that the dominant form of nitrogen in the solar nebula may have been N_2. On the other hand, kinetic inhibition to the conversion of N_2 to NH_3 (and CO to CH_4) must have been minimal in the nebulae of the outer planets because of the high pressures prevalent there. This led Prinn and Fegley (1981) to conclude that the form of C and N were, respectively, CH_4 and NH_3 in the circumplanetary nebula, including its cooler parts where Titan was formed. Incorporation of small amounts of CO and N_2 containing ices into some of the satellites of the outer planets is not precluded by this hypothesis. It is important to recognize that even if for some reason N_2 was the dominant form of nitrogen in the nebula, Titan's atmosphere could not have been the result of direct capture of N_2 during the accretionary phase. This becomes abundantly clear when one examines the ratio of ^{20}Ne to N_2. A directly captured atmosphere should exhibit the solar ratio of the elements, or $Ne/N_2 \approx 2$. In fact, neon has not even been detected on Titan; the upper limit of Ne/N_2 from spectroscopic considerations is 1% (Table 7.9). Furthermore, the atmospheric mean molecular weight is close to 28, which again argues against any significant proportion of neon in the atmosphere.

Clathrate-Hydrate Model

Miller (1961, 1973) suggested that the bulk of the mass of Saturn's satellites is contained in a methane hydrate, $CH_4 \cdot 7 H_2O$. Expanding on Miller's hypothesis, Owen (1982) proposed that the above-mentioned difficulty with neon could be circumvented, if instead of capturing N_2 directly, it was incorporated into a "clathrate-hydrate", and later released. A clathrate-hydrate is a lattice structure in which the host is made up of water-ice, while the guest molecules could be N_2, CH_4, CO, CO_2, Ar, and a whole lot of other cosmic constituents which would occupy the spaces of the water-ice lattice. Methane hydrate forms more easily and goes to completion at ~60 K. The vapor pressure of neon is 40 bar at 60 K, therefore there is no possibility of a neon-clathrate formation. The clathrate-hydrates, along with the silicates from the nebula, accrete to form Titan. Collisional and frictional heating, and later radiogenic heating, would liberate most of the volatiles which were incorporated in the ices. The nitrogen gas so released is stable in the atmosphere, except for a small loss due to escape. This scenario could explain the present nitrogen atmosphere of Titan. It does not necessarily require primordial Titan to have been warmer than today; however, it encounters the following difficulties:

a) If it is correct that nitrogen in the nebula was in the form of N_2, then the ratio of CH_4 to N_2 would be solar, or ~10. Such a large reservoir of methane ocean or aqueous solution would remove the bulk of the nitrogen from the atmosphere, as N_2 is readily soluble in methane. Replacing CH_4 by CO in the clathrate could permit more N_2 to stay in the atmosphere, as CO does not easily form clathrates. If it were indeed the case, nearly equal abundance of CO and N_2 would be found in the atmosphere. The measured CO/N_2 is close to 0.5%.

b) The devolatilization of the clathrate-hydrates should release argon in the solar ratio to N_2, i.e., $N_2/Ar \approx 11$. As discussed earlier, argon has neither been detected spectroscopically, nor is it required to explain the radio data from Voyager.

The above-mentioned difficulties with the clathrate model can be surmounted to a large extent if primordial Titan is assumed hot due to accretional heating, and the nitrogen in the nebula is in the form of NH_3, rather than N_2 (Hunten 1978; Atreya et al. 1978). According to a scenario presented by Lunine and Stevenson (1982), the photolysis of NH_3 resulted in modest amounts of N_2. This N_2, along with CH_4 and other cosmically abundant gases, formed a primordial atmosphere above a water-ammonia ocean. On cooling and condensation, a clathrate-hydrate formed which removed large amounts of atmospheric CH_4 (CH_4-clathrate forms more easily), but not much of N_2. Once the temperature dropped below 200 K, more N_2 can be removed, but at this temperature, the atmospheric vapor pressure of N_2 is 1 bar, and of CH_4 0.1 bar. Most of the water would form a thick ice stratum, which would prevent additional clathrate-hydrate formation, since aqueous solutions are almost necessary for clathrate formation. There must, however, be some mechanism for continuously supplying CH_4 to the atmosphere as the present atmosphere contains CH_4 and its photolysis products. Decomposition of clathrates at higher pressures of 30 – 40 bar in the interior of Titan is likely, and, in fact, necessary to recycle some of the CH_4 back to the atmosphere.

The less than solar abundance of CO and Ar can also be explained if the surface is indeed covered with liquid ethane. Such an ocean would trap large quantities of molecular and noble gases from the atmosphere. The thermodynamic properties of the proposed clathrate-hydrates are poorly understood, any quantitative model suffers from large uncertainties.

NH$_3$ Photolysis

The ammonia in the solar and circumplanetary nebula at Saturn, from which Titan was accreted, could also have been a direct source of Titan's atmospheric nitrogen. Lewis (1972) first suggested that Titan's atmosphere may be composed of N_2 supplied by NH_3 photolysis. The atmospheric model of Hunten (1978) also assumed N_2 as the major constituent. In a pre-Voyager model of NH_3 photochemistry, Atreya et al. (1978) presented many scenarios which can explain the present-day N_2 abundance on Titan (see Fig. 7.21). Ammonia has not been detected on Titan (Owen et al. 1977, place an upper limit of 30 cm-atm above the clouds). This is, however, not surprising, since the vapor pressure of NH_3 at the present surface temperature of Titan (94 K) is 2×10^{-8} mmHg, giving $NH_3/N_2 = 0.02$ ppb. Therefore, the NH_3 abundance is too low for detection, even if a continuous source of NH_3 existed on Titan, Moreover, detection of species below the clouds has not been possible so far. In the early stages of Titan's formation, however, accretional heating, radiogenic heating, heating due to micro-meteoroid impact, and possibly greenhouse effect could have resulted in a much warmer temperature. In fact, models of Lunine and Stevenson (1982) propose temperatures of 300 – 400 K in the early stages, although the depth at which these temperatures are reached is uncertain. In another model, Pollack (1973) finds that after outgassing began, and the surface pressure approached 0.5 bar of CH_4, a temperature of 150 K could be reached due to the greenhouse effect alone. (The present CH_4 surface pressure, <100 mb, is much less than the above-mentioned value, since the bulk of CH_4 may have been irreversibly converted to the heavier hydrocarbons, or incorporated into the ocean.) In any event, possibilities of much warmer Titan abound in its primordial stage.

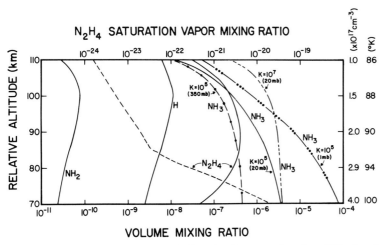

Fig. 7.21. Several scenarios for the evolution of an N_2 atmosphere from NH_3 photolysis on Titan. The altitude scale is arbitrary, however the important atmospheric parameters (density and temperature) are indicated on the *right*. *Solid lines* give NH_3, N_2H_4, NH_2, and H mixing ratios with an assumed cloud top pressure of 20 mb (due to CH_4 alone, in the past), and $K = 10^5\ cm^2\ s^{-1}$. NH_3 mixing ratios are shown also for (i) $K = 10^5\ cm^2\ s^{-1}$ and cloud top pressures of 1 mb and 350 mb, and (ii) $K = 10^7\ cm^2\ s^{-1}$ and cloud pressure of 20 mb. (Atreya et al. 1978)

Ammonia is a highly volatile compound. Even at the low temperatures prevalent in the Jovian troposphere, its photolysis proceeds rapidly. On primordial Titan, however, one must contend with the possibility that water might also have been present along with ammonia during Titan's warm phase, (presence of CH_4 does not inhibit N_2 production from the NH_3 photolysis, as discussed in the context of Jupiter in Chap. 5 on Photochemistry). Chemical reactions between NH_3 and H_2O and between their products are quite rapid, so that conversion of NH_3 to N_2 through photolysis will be prevented. At temperatures below 200 K, however, the saturation vapor pressures of NH_3 and H_2O begin to diverge considerably from each other, giving rise to a possibility where NH_3 photolysis could be considered in isolation. At the melting point of NH_3 (195.3 K), e.g., the NH_3 vapor pressure (~ 45 mmHg) is nearly 10^5 times greater than the H_2O vapor pressure (5.8×10^{-4} mmHg). In fact, NH_3 photolysis at 200 K and all the way down to 130 K results in the production of N_2. At the lower temperatures, however, the vapor pressure of a crucial intermediate product of NH_3 photolysis, hydrazine (N_2H_4), is much lower than its calculated partial pressure, so that N_2H_4 would condense and not produce N_2. Unlike Jupiter, however, N_2 cannot be recycled back into NH_3 on Titan since a deep and hot fluid interior does not exist. Without continuous outgassing, NH_3 would be irreversibly converted to small amounts of N_2 on Titan in a matter of a few thousand years.

Atreya et al. (1978) calculate that the NH_3 photolysis rate at Titan is $6.4 \times 10^{10}\ cm^{-2}\ s^{-1}$, assuming $K = 10^5\ cm^2\ s^{-1}$; and $T = 150$ K. Thus, it could take up to a hundred million years to accumulate the present surface pressure of N_2 on Titan. This seems too long for accreting Titan to maintain a temperature of >150 K. The time constant for the accumulation of the present N_2 atmosphere on Titan, however, may be considerably shorter, since the solar EUV flux in the past was presumably greater. IUE measurements of the pre-main sequence stars indicate that before the Sun reached the main sequence at an age of about 50 m.y., the solar flux shortward of 2000 Å may have been 10^4 times the present value (Zahnle and Walker 1982). The decrease in the solar activity is caused by the loss of angular momentum to the solar wind. This increased solar EUV flux has profound implications for the evolution of Titan's atmosphere. The time required for accumulating 1.5 bar of N_2 on Titan would be shortened to less than 50,000 years. Allowing for nitrogen escape, particularly when Titan is inside Saturn's magnetosphere and ions are generated, the time needed would still be less than 100,000 years. The correlation between the solar EUV flux and the time constant for atmospheric evolution is not expected to be a linear one as just assumed. Nevertheless, it is safe to say that the time scale will be short enough for an excellent possibility to exist for the accumulation of the present levels of N_2 on Titan before any clouds formed, and before the temperature dropped so low that the NH_3 to N_2 conversion was choked off due to the N_2H_4 condensation. The scenario just outlined for the generation of a massive atmosphere of nitrogen on Titan is tantamount to a cataclysmic event, occurring early in the accretionary phrase.

In summary, the most plausible scenario for the evolution of the present atmospheric composition of Titan involves some combination of ammonia photolysis and clathrate-hydrate devolatilization on primordial Titan, which is postulated to be considerably warmer than it is today.

7.5 Triton

The study of Triton is in its infancy because of the difficulties associated with observing small objects from Earth. Triton's angular diameter is only $0.15''$ as viewed from the Earth. Both reflection as well as absorption spectroscopy techniques are only marginally suitable for such an object which receives only 1/1000 of the solar energy compared to the Earth, and whose surface and atmospheric temperatures may be as cold as 50 to 60 K. Early telescopic observations showed that Triton is in a retrograde orbit with $160°$ inclination relative to the prograde Neptune's equator. As a consequence, strong tidal friction between Triton and Neptune would cause decay of the satellite's orbit until it is gravitationally captured by Neptune. In that circumstance, it would either crash into or be broken up in the vicinity of Neptune, in a matter of $10-100$ m.y. (McCord 1966; Trafton 1974). The following paragraphs present some speculation on Triton's atmosphere and chemistry based on the presently available ground-based data. Voyager 2 observations of Triton — planned for August, 1989, at an encounter distance of 5000 km — are expect-

Table 7.11. Atmospheric and surface composition of Triton

Constituent	Detection wavelength	Abundance	Reference
Nitrogen (liquid), N_2	(2-0) Band at 2.16 μm, tentative	10's of cm deep at the surface	1
Nitrogen (gas), N_2	(2-0) Band at 2.16 μm, tentative	$0.13 - 0.3$ atm surface pressure	1
Water frost, H_2O	Speculated from above spectra	?	1
Dark hydrocarbon (formed in CH_4 photochemistry?)	Speculated from above spectra	?	1
Methane, CH_4	2.3 μm (?); 0.89 μm	7 ± 3 m-atm; 50 m-atm	2 3

[1] Cruikshank et al. (1984); [2] Cruikshank and Silvaggio (1979); [3] Benner et al. (1978).

ed to provide useful data on its physical, orbital, atmospheric, and charged particle characteristics.

From the near infrared absorptions observed between 0.8 and 2.5 μm, Cruikshank and Apt (1984) have concluded the presence of CH_4 on Triton, probably in the solid state. Gaseous methane in vapor pressure equilibrium is expected. In addition, a relatively broad feature at 2.16 μm can be best interpreted by attributing it to liquid nitrogen. A layer at least tens of cm deep over much of the surface is estimated (Cruikshank et al. 1984). Pressure-induced absorption in gaseous N_2 occurs also at the same wavelength as liquid nitrogen, i.e., 2.16 μm. Cruikshank et al. conclude from the models they used to fit their data that the surface pressure of gaseous N_2 is 130 to 300 mb. This range actually represents approximately the N_2 vapor pressures in equilibrium with pure liquid nitrogen at 64 K and 67 K, respectively, (vapor pressure relations for N_2 are given in Chap. 3 on Cloud Structure). The actual vapor pressure, however, could be lower, as the surface most likely contains impurities. In fact, fine-grained water frost and dark hydrocarbon solids are needed to explain the shape of the continuum slope adjacent to the observed 2.1 and 2.4 μm features (Cruikshank et al. 1984). The current understanding of the atmospheric and surface composition of Triton based on the observations of Cruikshank et al. is summarized in Table 7.11. Here it is important to note that the identification of the 2.1 μm feature with liquid nitrogen should be regarded as tentative. Higher spectral resolution observations of Triton when it was in its eastern elongation (Rieke et al. 1985) do not confirm the above conclusion of Cruikshank et al. (1984). Either liquid nitrogen does not exist on Triton, or it is nonuniformly distributed. It is also possible that the 2.1 μm feature is much too broad for it to be scanned by the high resolution spectrometer of Rieke et al. (1985). In fact, the observations of Cruikshank et al.

are equally consistent with a nonuniform distribution of liquid nitrogen, perhaps in the form of lakes, ponds, or puddles.

The equilibrium temperature at the subsolar point depends critically on the bolometric albedo, A, a quantity known very poorly for Triton. Assuming $A = 0.4$ (after Morrison et al. 1982), Cruikshank et al. calculate the subsolar temperature to be 64 K (for bolometric emissivity, $\varepsilon_b = 1$), or 67 K (for $\varepsilon_b = 0.8$). The disc average temperature is 57 K. The assumptions about radius (here, 1750 km) and density (1.3 gm cm^{-3}) can also affect the surface temperature. None of these quantities is known with a high degree of precision. The dry adiabatic lapse rate at the surface turns out to be 0.6 K km^{-1}, and is not changed by van der Waals correction terms. The radiative equilibrium time constant at the surface is calculated to be 5×10^9 s, which is comparable to the seasonal cycle of 4×10^9 s. Thus there is a good probability of seasonal changes on Triton. The slow rotation rate of the satellite would tend to damp out large latitudinal changes in temperature.

Liquid nitrogen in contact with CH$_4$-vapor is somewhat analogous to water-CO$_2$ system in the Earth's atmosphere. CH$_4$ would dissolve in liquid N$_2$ according to the following relation (Fastovsky and Krestinsky 1941):

$$\log N = 1.36576 - 120.48/T, \tag{7.13}$$

where N is the solubility in mol-%.

At the subsolar temperature of 64 K, the solubility of CH$_4$ in liquid N$_2$ will be 30%. It is a strong function of the temperature, e.g., it is less than 20% for the disc average temperature of 57 K, and 40% with $T = 67$ K, the subsolar temperature with $\varepsilon_b = 0.8$. The effect of doping liquid N$_2$ with CH$_4$ and other impurities such as water frost, neon, etc., will be to lower its melting point, and thus enhance the probability of liquid nitrogen ponds on the surface.

The CH$_4$ chemistry in liquid N$_2$ proceeds slowly. Over geologic time, however, numerous complex molecules such as nitrites, azides, etc. could form. In the atmosphere, there is plenty of CH$_4$ in the two extreme cases cited in Table 7.11. Approximately 10^{22} to 10^{23} cm^{-2} column abundance of CH$_4$ is available at the surface (note that the abundances of CH$_4$ in Table 7.11 roughly correspond to its saturation vapor pressures at the above-mentioned subsolar temperatures). For unit optical depth in CH$_4$, only 5×10^{16} cm^{-2} column abundance is needed. Thus, photolysis of CH$_4$, in principle can occur high in the atmosphere, perhaps 5 to 10 scale heights above the surface. The scale height at the surface is 30 km, and if the temperature drops according to the dry adiabatic lapse rate in the troposphere (0.6 K km^{-1}), then the saturation vapor pressure of CH$_4$ at the homopause would control its density there. Photolysis of CH$_4$ is still expected to take place above the tropopause, although the photolysis rates would be reduced by a factor of 2 over the values at Titan. Somewhat smaller production of C$_2$H$_2$ and C$_2$H$_6$ would occur. However, the gas phase abundances of C$_2$H$_2$ and C$_2$H$_6$ are expected to be extremely low because of their condensation near and above the tropopause, as on Uranus and Neptune (see Chap. 5 on Photochemistry).

The nitrogen chemistry on Titan is controlled by the magnetospheric charged particle input, since the solar EUV fluxes are low. At Triton, the solar fluxes are even less effective in triggering the onset of nitrogen chemistry. There is little likelihood of large magnetospheric energetic particle input at Triton, since the IUE data on the Neptune Lyman-α do not indicate the presence of a magnetosphere. An upper limit of $300\,R$ has been placed on the Lyman-α emission from Neptune (J. T. Clarke 1985, personal communication). The emission rate is low enough to be explained by resonance scattering of the solar Lyman-α photons by the photochemically produced hydrogen atoms. Additional hydrogen atom production by charged particle dissociation of H_2, or direct charged particle excitation is not required. In the absence of large energy input from charged particles at Triton, the production of NI, N^+, and N_2^+ would be quite small. Production of organics of $C-N-H$ molecules would be inhibited, despite the fact that CH_4/N_2 ($\approx 10^{-3}$) ratio is respectable. Again, the low temperatures on Triton would prevent practically any organics or complex hydrocarbons produced in the $C-N-H$ chemistry from existing in the vapor phase.

The ammonia photolysis and the clathrate-hydrate devolatilization scenarios discussed for explaining Titan's nitrogen atmosphere are also applicable to Triton. In addition to the heating mechanisms proposed for Titan, tidal friction can contribute to Triton's heating. The tidal heating effects are, however, expected to be small and latitude-dependent.

In summary, the presence of an atmosphere on Triton is virtually guaranteed. The question is whether it is composed of just CH_4, or N_2 as well. In the former instance, methane photochemistry similar to that on Uranus and Neptune is expected to proceed. If the latter were to hold, Triton's atmosphere would look somewhat like Titan's. The surface may be composed of liquid nitrogen with methane dissolved in it, although a nonhomogeneous mixture of solid N_2 and CH_4 crystals cannot be ruled out from purely thermodynamical considerations. The Voyager observations at the infrared, radio, and ultraviolet wavelengths will unravel the mystery surrounding this unique object in our solar system.

7.6 Outstanding Issues

Are the volcanoes on Io always as active as at the time of the Voyager 1 and Voyager 2 observations, or are they episodic? Is the SO_2 atmosphere on Io a transient one driven by the volcanic activity? Is the atmosphere close to the "thick" or a "thin" pressure limit? What is responsible for the nightside ionization on Io? What is the ionospheric composition? How does the material from Io's surface and the atmosphere modify the atmosphere of Jupiter? Do tenuous atmospheres exist on the other Galilean moons as well? What are the clouds of Titan made of? What is the atmospheric composition below the cold-trap? What is Titan's surface composition? How does its atmosphere change with seasons, both at the low as well as high latitudes? What is the composition and the altitude profile of the ions on Titan? How does it change with the change in the boundary and the intensity of Saturn's magnetosphere? What are the abundances and the isotopic ratios of the noble gases in Titan's homosphere? How warm were Titan's surface and the lower troposphere during its early phase of accretion, and for how long? For Triton, even the basic information on its atmospheric thermal structure and the composition is lacking. Long-term observations of Io and the other Galilean satellites from the Galileo Orbiter, of Triton from Voyager, and of Titan from Cassini/Saturn Orbiter and Titan Probe will be extremely valuable in addressing many of the above-mentioned issues. Observations from earth orbit (e.g., with Hubble Space Telescope) can also yield useful data. Questions of the evolution of an atmosphere on Titan (and possibly Triton) would require further theoretical modeling, including effects of charged particle stripping of the atmosphere and, possibly clouds during the primordial phase. Laboratory data on the vapor pressures, latent heats, heat capacities and solubilities relevant for multi-phase binary and tertiary mixtures will also be needed. All these data will be meaningful only if collected under conditions of Titan's temperature, pressure, and composition.

General Appendix A.1. Characteristics of the outer planets and comparison with the Earth[a]

Characteristics	Planet					
	Earth	Jupiter	Saturn	Uranus	Neptune	Pluto
Mean distance from Sun [AU][b]	1.000	5.202	9.554	19.218	30.109	39.44
Mass, M [10^{24} kg]	5.975	1.900	569	87	103	0.0113(?)
Equatorial radius, R_E [km][c]	6,378	71,492	60,268	25,550	24,300(?)	1,500 ± 200(?)
Dynamic oblateness	0.0034	0.0637	0.102	0.024(?)	0.0266(?)	?
Mean density, ρ [gm cm^{-3}]	5.52	1.314	0.69	1.25	1.66(?)	0.8(?)
Equatorial surface gravity, g [cm s^{-2}]	978	2288	904	865	1100(?)	40(?)
Escape velocity [km s^{-1}]	11	57	33	20	23(?)	2(?)
Equatorial rotation period [h]	23.9345	9.841	10.233	17.29 ± 0.1	15.8(?)	153.3(?)
Orbital sidereal period [yrs]	1.000	11.8623	29.458	84.01	164.79	248.5
Inclination of equator to orbit [degrees]	23.44	3.08	29	97.92	28.8	≥50(?)
Magnetic dipole moment [Gauss-cm^3]	8.06×10^{25}	1.6×10^{30}	4.7×10^{28}	3.84×10^{27}	?	?
Dipole equatorial magnetic field [Gauss]	0.31	4.28	0.21	0.23	?	?
Tilt of magnetic axis relative to spin axis [deg]	+11.7	−9.6	0	−60	?	?

[a] Entries marked (?) are uncertain.
[b] 1 AU = 1.496×10^{13} cm.
[c] Radii of the outer planets are given above the 1-bar pressure level because of the lack of solid surface for altitude reference.

General Appendix A.2. Characteristics of the large satellites of the outer planets and comparison with Earth's moon[a]

Characteristic	Satellite (of planet)						
	(Earth's)	(Jupiter's)				(Saturn's)	(Neptune's)
	Moon	Io	Europa	Ganymede	Callisto	Titan	Triton
Mean distance from planet center (R_{planet})	60.27 R_E	5.78 R_J	9.40 R_J	14.99 R_J	26.33 R_J	20.25 R_S	14.61 R_N
Mass, M [10^{22} kg]	7.35	8.92	4.87	14.90	10.75	13.5	3.0
Equatorial radius, R_e [km]	1738	1815 ± 5	1569 ± 10	2631 ± 10	2400 ± 10	2575 ± 0.5	1750(?)
Mean density, ρ [gm cm^{-3}]	3.341	3.55	3.04	1.93	1.83	1.88	1.3
Equatorial surface gravity, g [cm s^{-2}]	162.3	180.7	132.0	143.6	124.5	135.4	64
Sidereal period, [days]	27.321	1.769	3.551	7.155	16.689	15.945	5.877
Orbital inclination [degrees]	23	0.027	0.468	0.183	0.253	0.33	160
Orbital eccentricity	0.055	0.000	0.000	0.001	0.007	0.029	0.00
Albedo, A	0.067	0.6	0.6	0.4	0.2	0.2	0.4
Escape velocity [km s^{-1}]	2.38	2.56	2.04	2.75	2.44	2.64	1.50

[a] Jupiter has 16 known moons; Saturn, 17; Uranus, 15; Neptune, 2; and Pluto, 1. In addition, Jupiter, Saturn and Uranus all have rings, while Neptune is suspected to have them (See Beatty et al. 1982; Burns 1977; Morrison 1982; and Science 1986 for characteristics of the satellites not listed in the above table).

General Appendix A.3. Useful numerical constants

Gravitational constant	$G = 6.670 \times 10^{-8}$ dyne cm^2 gm^{-2}
Velocity of light	$c = 2.99793 \times 10^{10}$ cm s^{-1}
Planck's constant	$h = 6.62517 \times 10^{-27}$ erg s
Electronic charge	$e = 4.80286 \times 10^{-10}$ esu; 1.60206×10^{-20} emu
Electron rest mass	$m_e = 9.1083 \times 10^{-28}$ gm
Proton rest mass	$m_p = 1.67239 \times 10^{-24}$ gm
Classical electron radius	$r_0 = 2.8178 \times 10^{-13}$ cm
Fine structure constant	$\alpha = 1/137$
Avogadro's constant	$N_A = 6.02486 \times 10^{23}$ molecules mol^{-1}
Boltzmann constant	$k = 1.38044 \times 10^{-16}$ erg K^{-1}
Gas constant per mole	$R = 8.31696 \times 10^7$ erg mol^{-1} K^{-1}
Loschmidt's number	$L = 2.68719 \times 10^{19}$ molecules cm^{-3}
Stefan-Boltzmann constant	$\sigma = 0.56687 \times 10^{-4}$ erg cm^{-2} s^{-1} deg^{-4}

References

Adams NG, Smith D (1976) Production distributions for some ion-molecule reactions. J Phys B 9:1439–1451

Aitken DK, Smith C, Roche PF, Orton GS, Caldwell J, Snyder R (1986) The spectra of Uranus and Neptune at 8–14 and 17–23 μm. Preprint

Akimoto H, Obi K, Tanaka I (1965) Primary process in the photolysis of ethane at 1236 Å. J Chem Phys 42:3864–3868

Allen M, Yung YL (1985) A simple photochemical model for forming benzene in the Jovian atmosphere. Bull Am Astron Soc 17:710

Allen M, Pinto JP, Yung YL (1980) Aerosol photochemistry and variations related to the sunspot cycle. Astrophys J Lett 242:L125–L128

Ames WF (1969) Numerical methods for partial differential equations. Barnes and Noble, New York

Appleby J (1980) PhD Thesis, SUNY at Stony Brook, New York

Appleby JF (1986) Radiative-convective equilibrium models of Uranus and Neptune. Icarus 65:383–405

Atreya SK (1981) Measurement of minor species (H_2, Cl, O_3, NO) in the earth's atmosphere by the stellar occultation technique. In: Atreya SK, Caldwell JJ (eds) Planetary Aeronomy and Astronomy. Adv Space Res 1:127–141

Atreya SK (1982) Eddy diffusion coefficient on Saturn. Planet Space Sci 30:849–854

Atreya SK (1984a) Aeronomy. In: Bergstralh JT (ed) Uranus and Neptune. NASA Conf Publ 2330, US Government Printing Office, Wash DC, pp 55–88

Atreya SK (1984b) Modification of planetary atmospheres by material from the rings. Adv Space Res 4:31–40

Atreya SK, Donahue TM (1975a) Ionospheric models of Saturn, Uranus and Neptune. Icarus 24:358–362

Atreya SK, Donahue TM (1975b) The role of hydrocarbon in the ionosphere of the outer planets. Icarus 25:335–338

Atreya SK, Donahue TM (1976) Model ionospheres of Jupiter. In: Gehrels T (ed) Jupiter. Univ Arizona Press, Tucson, pp 304–318

Atreya SK, Donahue TM (1979) Models of the Jovian upper atmosphere. Rev Geophys Space Phys 17:388–396

Atreya SK, Ponthieu JJ (1983) Photolysis of methane and the ionosphere of Uranus. Planet Space Sci 33:939–944

Atreya SK, Romani PN (1985) Photochemistry and clouds of Jupiter, Saturn and Uranus. In: Hunt GE (ed) Planetary Meteorology. Cambridge Univ Press, pp 17–68

Atreya SK, Waite JH Jr (1981) Saturn ionosphere: theoretical interpretation. Nature 292: 682–683

Atreya SK, Donahue TM, McElroy MB (1974) Jupiter's ionosphere: prospects for Pioneer 10. Science 184:154–156

Atreya SK, Donahue TM, Sharp WE, Wasser B, Drake JF, Riegler GR (1976) Ultraviolet stellar occultation measurement of the H_2 and O densities near 100 km in the earth's atmosphere. Geophys Res Lett 3:607–610

Atreya SK, Yung YL, Donahue TM, Barker ES (1977a) Search for Jovian auroral hot spots. Astrophys J 218:L83–L87

Atreya SK, Donahue TM, Kuhn WR (1977b) The distribution of ammonia and its photochemical products on Jupiter. Icarus 31:348 – 355

Atreya SK, Donahue TM, Kuhn WR (1978) Evolution of a nitrogen atmosphere on Titan. Science 201:611 – 613

Atreya SK, Donahue TM, Sandel BR, Broadfoot AL, Smith GR (1979a) Jovian upper atmospheric temperature measurement by the Voyager 1 UV spectrometer. Geophys Res Lett 6:795 – 798

Atreya SK, Donahue TM, Waite JH Jr (1979b) An interpretation of the Voyager measurement of Jovian electron density profiles. Nature 280:795 – 796

Atreya SK, Kuhn WR, Donahue TM (1980) Saturn: Tropospheric ammonia and nitrogen. Geophys Res Lett 7:474 – 476

Atreya SK, Donahue TM, Festou MC (1981) Jupiter: structure and composition of the upper atmosphere. Astrophys J 247:L43 – L47

Atreya SK, Festou MC, Donahue TM, Kerr RB, Barker ES, Cochran WD, Bertaux JL, Upson WL II (1982) Copernicus measurement of the Jovian Lyman-alpha emission and its aeronomical significance. Astrophys J 262:377 – 387

Atreya SK, Waite JH, Donahue TM, Nagy AF, McConnell JC (1984) Theory, measurements and models of the upper atmosphere and ionosphere of Saturn. In: Gehrels E (ed) Saturn. Univ Arizona Press, Tucson, pp 239 – 277

Auerbach D, Cacak R, Candano R, Gaily TD, Keyser CJ, McGowan JW, Mitchell JBA, Wilk SFJ (1977) Merged electron-ion beam experiments. I. Methods and measurements of $(e\text{-}H_2^+)$ and $(e\text{-}H_3^+)$ dissociative recombination cross sections. J Phys B 10:3797 – 3820

Axel L (1972) Inhomogeneous models of the atmosphere of Jupiter. Astrophys J 173:451 – 468

Back RA, Griffiths DWL (1967) Flash photolysis of ethylene. J Chem Phys 46:4839 – 4843

Bagenal F, Sullivan JD (1981) Direct plasma measurements in the Io torus and inner magnetosphere of Jupiter. J Geophys Res 86:8447 – 8466

Banks PM, Kockarts G (1973) Aeronomy, vol A and B. Academic Press, New York

Barker ES, Cazes S, Emerich C, Vidal-Madjar A, Owen T (1980) Lyman-alpha observations in the vicinity of Saturn with Copernicus. Astrophys J 242:383 – 394

Barshay SS, Lewis JS (1978) Chemical structure of the deep atmosphere of Jupiter. Icarus 33:593 – 607

Bartholdi P, Owen F (1972) The occultation of beta scorpii by Jupiter and Io II. Astron J 77:60 – 65

Bates DR, Dalgarno A (1962) Electronic recombination. In: Atomic and molecular processes. Academic Press, New York, pp 245 – 271

Bauer SJ (1973) Physics of planetary ionospheres. Springer, Berlin Heidelberg New York

Baulch DL, Drysdale DD (1976) Evaluated kinetic data for high temperature reactions. The Butterworth Group, Univ Leeds

Beatty JK, O'Leary B, Chaikin A (1982) The new solar system. Cambridge Univ Press

Beer R (1975) Detection of carbon monoxide on Jupiter. Astrophys J 200:L167 – L170

Beer R, Taylor FW (1973a) The abundance of CH_3D and the D/H ratio in Jupiter. Astrophys J 179:309 – 327

Beer R, Taylor FW (1973b) The equilibrium of deuterium in the Jovian atmosphere. Astrophys J 182:1131 – 1132

Beer R, Taylor FW (1978) The abundance of carbon monoxide in Jupiter. Astrophys J 221:1100 – 1109

Beer R, Taylor FW (1979) Phosphine absorption in the 5 μm window of Jupiter. Icarus 40:189 – 192

Belton MJS (1982) An interpretation of the near-ultraviolet absorption spectrum of SO_2: implications for Venus, Io and laboratory measurements. Icarus 52:149 – 165

Benner DC, Fink U, Cromwell RH (1978) Image tube spectra of Pluto and Triton from 6800 to 9000 Å. Icarus 36:82 – 91

Bergstralh JT, Baines KH (1984) Properties of the upper troposphere of Uranus and Neptune derived from observations at visible to near-infrared wavelengths. In: Bergstralh J (ed) Uranus and Neptune. NASA Conf Publ 2330:179 – 212

Bertaux JL, Belton MJS (1979) Evidence of SO_2 on Io from UV observations. Nature 282:813–815

Bertaux JL, Kockarts G (1983) Distribution of molecular hydrogen in the atmosphere of Titan. J Geophys Res 88:8716–8720

Bertaux JL, Festou M, Barker E, Jenkins E (1980) Copernicus measurements of the Lyman Alpha albedo of Jupiter. Astrophys J 238:1152–1159

Bjoraker GL (1985) The gas composition and vertical cloud structure of Jupiter's troposphere derived from five micron spectroscopic observations. Thesis, Univ Arizona

Bridge HS, Belcher JW, Lazarus AJ, Sullivan JD, McNutt RL, Bagenal F, Scudder JD, Sittler ES, Siscoe GL, Vasyliunas VM, Goertz CK, Yeates CM (1979) Plasma observations near Jupiter: initial results from Voyager 1. Science 204:987–991

Bridge HS, Belcher JW, Lazarus AJ, Olbert S, Sullivan JD, Bagenal F, Gazis PR, Hartle RE, Ogilvie KW, Scudder JD, Sittler EC, Eviatar A, Siscoe G, Goertz C, Vasyliunas V (1981) Plasma observations near Saturn: initial results from Voyager 1. Science 212:217–224

Broadfoot AL, Belton MJS, Takacs PZ, Sandel BR, Shemansky DE, Holberg JB, Ajello JM, Atreya SK, Donahue TM, Moos HW, Bertaux JL, Blamont JE, Strobel DF, McConnell JC, Dalgarno A, Goody R, McElroy MB (1979) Extreme ultraviolet observations from Voyager 1 encounter with Jupiter. Science 204:979–982

Broadfoot AL, Sandel BR, Shemansky DE, Holberg JB, Smith GR, Strobel DF, Kumar S, McConnell JC, Hunten DM, Atreya SK, Donahue TM, Moos HW, Bertaux JL, Blamont JE, Pomphray RB, Linick S (1981a) Extreme ultraviolet observations from Voyager 1 encounter with Saturn. Science 212:206–211

Broadfoot AL, Sandel BR, Shemansky DE, McConnell JC, Atreya SK, Donahue TM, Strobel DF, Bertaux JL (1981b) Overview of the ultraviolet spectrometry results through Jupiter encounters. J Geophys Res 86:8259–8284

Broadfoot AL, Herbert F, Holberg J, Hunten D, Sandel B, Shemansky D, Smith G, Yelle R, Kumar S, Strobel D, Moos H, Atreya S, Donahue T, Bertaux J, Blamont J, McConnell J, Dessler A, Linnick S, Springer R (1986) Ultraviolet spectrometer observations of Uranus. Science 233:74–79

Brown RA (1974) Optical line emissions from Io. In: Woszczyk A, Iwaniszewska C (eds) Exploration of the planetary system. Reidel, Dordrecht, Holland, pp 527–531

Brown RA (1981) The Jupiter hot plasma torus: Observed electron temperature and energy flow. Astrophys J 244:1072–1080

Brown RL, Laufer AH (1981) Calculation of activation energies for hydrogen-atom abstractions by radicals containing carbon triple bonds. J Phys Chem 85:3826–3828

Brown WL, Lanzerotti JL, Poate JM, Augustyniak WM (1978) Sputtering of ice by MeV light ions. Phys Rev Lett 40:1027–1030

Browning R, Fryar J (1973) Dissociative photoionization of H_2 and D_2 through the 1s σ_g ionic state. Proc Phys Soc Lond 6:364–371

Burns JA (1977) Planetary satellites. Univ Arizona Press

Burns JA, Showalter MR, Cuzzi JN, Durisen RH (1983) Saturn's electrostatic discharges: could lightning be the cause? Icarus 54:280–295

Butler JE, Fleming JW, Gross LP, Lin MC (1981) Kinetics of CH radical reactions with selected molecules at room temperature. Chem Phys 56:355–365

Butterworth PS (1980) A new upper limit to the global sulfur dioxide abundance on Io. Paper presented at the Satellites of Jupiter, Symp 57. Int Astron Union, Kailua-Kona, Hawaii, May 13–16

Caldwell J (1977) Ultraviolet observations of Mars and Saturn by the TDIA and OAO-2 Satellites. Icarus 32:190–209

Caldwell J, Tokunaga AT, Gillett FC (1980) Possible infrared aurorae on Jupiter. Icarus 44:667–675

Caldwell J, Tokunaga AT, Orton GS (1982) Further observations of 8 µm polar brightenings of Jupiter. Icarus 53:133–140

Caldwell J, Wagener R, Owen T, Combes M, Encrenaz T (1984) Ultraviolet observations of Uranus and Neptune below 3000 Å. Uranus and Neptune. NASA Conf Publ 2330:157–178

Callear AB, Metcalfe MP (1976) Oscillator strengths of the bands of the $\tilde{B}\,^2A_1$-$\tilde{X}\,^1A_2$ system of CD_3 and a spectroscopic measurement of the recombination rate. Comparison with CH_3. Chem Phys 14:275–284

Calvert JF, Pitts JN (1966) Photochemistry. Wiley, New York

Cameron AGW (1973) Abundances of the elements in the solar system. Space Sci Rev 15:121–146

Cameron AGW (1982) Elemental and Nuclidic abundances in the Solar System. In: Barnes CA, Clayton DD, Schramon DN (eds) Essays in nuclear astrophysics. Cambridge Univ Press

Capone LA, Dubach J, Whitten RC, Prasad SS (1979) Cosmic ray ionization of the Jovian atmosphere. Icarus 39:433–449

Carlson RW, Judge DL (1974) Pioneer 10 UV Photometer observations of the Jupiter encounter. J Geophys Res 79:3623–3633

Carlson RW, Bhattacharyya C, Smith BA, Johnson TV, Hidayat B, Smith SA, Taylor GE, O'Leary B, Brinkmann RT (1973) An atmosphere on Ganymede from its occultation of SAO18600 on 7 June 1972. Science 182:53–55

Cerceau F, Raulin F, Courtin R, Gautier D (1985) Infrared spectra of gaseous mononitriles: application to the atmosphere of Titan. Icarus 62:207–220

Chamberlain JW (1963) Planetary coronae and atmospheric evaporation. Planet Space Sci 11:901–960

Chamberlain JW (1978) Theory of planetary atmospheres. Academic Press, New York

Chamberlain JW, McElroy MB (1966) Martian atmosphere: the Mariner occultation experiments. Science 152:21–22

Chandrasekhar S (1960) Radiative transfer. Dover, New York

Chapman S, Cowling TG (1970) The mathematical theory of nonuniform gases. Cambridge Univ Press, London

Chen RH (1981) Studies of Jupiter's lower ionospheric layers. J Geophys Res 86:7792–7794

Chen RH (1983) Saturn's ionosphere. A corona of ice particles? In: Moon and planets, vol 28. Reidel, Dordrecht, pp 37–41

Cheng AF, Lanzerotti LJ, Pironello V (1982) Charged particle sputtering of ice surfaces in Saturn's magnetosphere. J Geophys Res 87:4567–4570

Clarke JT (1982) Detection of auroral H Ly-α emission from Uranus. Astrophys J 263: L105–L109

Clarke JT, Moos HW, Atreya SK, Lane AL (1980a) Observations from earth orbit and variability of the polar aurora on Jupiter. Astrophys J 241:L179–L182

Clarke JT, Weaver HA, Feldman PD, Moos HW, Fastie WG, Opal CB (1980b) Spatial imaging of hydrogen Lyman-α emission from Jupiter. Astrophys J 240:696–701

Clarke JT, Moos HW, Atreya SK, Lane AL (1981a) International ultraviolet explorer detection of bursts of hydrogen Lyman-alpha emission from Saturn. Nature 290:226–227

Clarke JT, Moos HW, Atreya SK, Lane AL (1981b) Observations of auroras on Jupiter. In: The universe in ultraviolet wavelengths: the first two years of IUE. NASA Conf Publ 2171, vol 45. US Government Printing Office, Wash DC

Clarke J, Durrance S, Atreya SK, Barnes A, Belcher J, Festou M, Imhoff C, Mihalov J, Moos W, Murthy J, Pradhan A, Skinner T (1986) Continued observations of the H Ly-alpha emission from Uranus. J Geophys Res 91:8771–8781

Cochran WD, Barker ES (1979) Variability of Lyman-alpha emission from Jupiter. Astrophys J 234:L151–L154

Colegrove FD, Johnson FS, Hanson WB (1966) Atmospheric composition in the lower thermosphere. J Geophys Res 71:2227–2236

Combes M, Maillard J, deBergh C (1977) Evidence for a telluric value of $^{12}C/^{13}C$ ratio in the atmosphere of Jupiter and Saturn. Astron Astrophys 61:531–537

Conklin EK, Ulich BL, Dickel JR (1977) 3 mm observations of Titan. Bull Am Astron Soc 9:471

Connerney JEP, Waite JH (1984) Wet model of Saturn's Ionosphere: Water from the Rings. Nature 312:136–138

Conrath BJ, Pirraglia JA (1983) Thermal structure of Saturn from Voyager infrared measurements: implications for atmospheric dynamics. Icarus 53:286–292

Conrath BJ, Gautier D, Hanel RA, Hornstein JS (1984) The helium abundance of Saturn from Voyager measurements. Astrophys J 282:807 – 815

Courtin R, Gautier D, Lacombe A (1978) On the thermal structure of Uranus from infrared measurements. Astron Astrophys 63:97 – 101

Courtin R, Gautier D, Lacombe A (1979) Indications of supersaturated stratospheric methane on Neptune from its atmospheric thermal profile. Icarus 37:236 – 248

Courtin R, Gautier D, Marten A, Kunde V (1983) The $^{12}C/^{13}C$ ratio in Jupiter from Voyager infrared investigation. Icarus 53:121 – 132

Courtin R, Gautier D, Marten A, Bézard B, Hanel R (1984) The composition of Saturn's atmosphere at Northern temperate latitudes from Voyager IRIS Spectra: NH_3, PH_3, C_2H_2, C_2H_6, CH_3D, CH_4 and the Saturnian D/H ratio. Astrophys J 287:899 – 916

Cravens TE (1974) Astrophysical applications of electron energy deposition in molecular hydrogen. Thesis, Harvard Univ

Cruikshank DP (1980) Infrared spectrum of Io, 2.8 – 5.2 microns. Icarus 41:240 – 245

Cruikshank DP, Apt J (1984) Methane on Triton: physical state and distribution. Icarus 58:306 – 311

Cruikshank DP, Silvaggio PM (1978) Methane atmospheres of Triton and Pluto. Bull Am Astron Soc 10:578

Cruikshank DP, Silvaggio PM (1979) Triton: a satellite with an atmosphere. Astrophys J 233:1016 – 1020

Cruikshank DP, Silvaggio PM (1980) The surface and atmosphere of Pluto. Icarus 41:96 – 102

Cruikshank DP, Brown RH, Clark RN (1984) Nitrogen on Triton. Icarus 58:293 – 305

Danielson RE, Caldwell JJ, Larach DR (1973) An inversion in the atmosphere of Titan. Icarus 20:437 – 443

Davis DD, Okabe H (1968) Determination of bond dissociation energies in hydrogen cyanide, cyanogen and cyanogen halides by the photodissociation method. J Chem Phys 49:5526 – 5531

DePater I, Massie ST (1985) Models of the millimeter-centimeter spectra of the giant planets. Icarus 62:143 – 171

Dessler AJ (1983) Coordinate systems. In: Dessler AJ (ed) Physics of the Jovian magnetosphere. Cambridge Univ Press, pp 498 – 504

Driscoll JN, Warneck P (1968) Primary processes in the photolysis of SO_2 at 1849 Å. J Phys Chem 72:3736 – 3739

Drossart P, Serabyn E, Lacy J, Atreya SK, Bézard B, Encrenaz T (1985) C_2H_2, C_2H_6 and polar infrared brightening on Jupiter. Bull Am Astron Soc 17:708

Drossart P, Bézard B, Atreya SK, Lacey J, Serabyn E, Tokunaga A, Encrenaz T (1986) Enhanced acetylene emission near the north pole of Jupiter. Icarus 66:610 – 618

Dunham E, Elliot JL, Gierasch PJ (1980) The upper atmosphere of Uranus: mean temperature and temperature variations. Astrophys J 235:274 – 284

Durrance ST, Moos HW (1982) Intense Lyman-α emission from Uranus. Nature 299:428 – 430

Elliot JL, Wasserman LH, Veverka J, Sagan C, Liller W (1974) The occultation of Beta Scorpii by Jupiter 2. The hydrogen and helium abundance in the Jovian atmosphere. Astrophys J 190:719 – 729

Encrenaz T, Combes M (1978) On the D/C ratio in the atmosphere of Uranus. Bull Am Soc 10:567 – 577

Encrenaz T, Owen T (1973) New observations of the hydrogen quadrupole lines on Saturn and Uranus. Astron Astrophys 28:119 – 124

Encrenaz T, Combes M, Zeau Y, Vapillon L, Berezne J (1975) A tentative identification of C_2H_4 in the spectrum of Saturn. Astron Astrophys 42:355 – 356

Encrenaz T, Combes M, Zeau Y (1978) The spectrum of Jupiter between 10 and 13 microns: an estimate of the Jovian $^{14}N/^{15}N$ ratio. Astron Astrophys 70:29 – 36

Encrenaz T, Combes M, Atreya SK, Romani PN, Fricke K, Moore V, Hunt G, Wagener R, Owen T, Caldwell J, Butterworth P (1986) A study of the upper atmosphere of Uranus using IUE. Astron Astrophys 162:317 – 322

Eshleman VR (1982) Argon in Titan's atmosphere? Science 217:200

Eshleman VR, Tyler GL, Anderson JD, Fjeldbo G, Levy GS, Wood GE, Croft TA (1977) Radio science investigations with Voyager. Space Sci Rev 21:207 – 232

Eshleman VR, Tyler GL, Wood GE, Lindal GF, Anderson JD, Levy GS, Croft TA (1979a) Radio science with Voyager 1 at Jupiter: preliminary profiles of the atmosphere and ionosphere. Science 204:976 – 978

Eshleman VR, Tyler GL, Wood GE, Lindal GF, Anderson JD, Levy GS, Croft TA (1979b) Radio science with Voyager at Jupiter: initial Voyager 2 results and a Voyager 1 measure of the Io torus. Science 206:959 – 962

Eshleman VR, Lindal GF, Tyler GL (1983) Is Titan wet or dry? Science 221:53 – 55

Esposito LW (1984) Sulfur dioxide: Episodic injection shows evidence for active Venus volcanism. Science 223:1072 – 1074

Evans DR, Warwick JW, Pearce JB, Carr TD, Schauble JJ (1981) Impulsive radio discharges near Saturn. Nature 292:716 – 718

Evans WFJ, Kerr JB (1982) Estimates of the amount of sulfur dioxide injected into the stratosphere by the explosive volcanic eruptions: El Chicon, Mystery Volcano, Mt St. Helens. Geophys Res Lett 10:1049 – 1051

Fair RW, Thrush BA (1967) Reaction rate for sulfur atoms. Faraday Discuss Chem Soc 44: 236

Fanale FP, Banerdt WB, Elson LS, Johnson TV, Zurek RW (1980) Io's surface: its phase composition and influence on Io's atmosphere and Jupiter's magnetosphere. Paper presented at IAU Colloquium 57. Int Astron Union, Kailua-Kona, Hawaii

Fastovsky VG, Krestinsky YA (1941) J Phys Chem 15:525 – 531 (in Russian)

Ferris JP, Benson R (1980) Diphosphine is an intermediate in the photolysis of phosphine to phosphorus and hydrogen. Nature 285:156 – 157

Festou MC, Atreya SK (1982) Voyager ultraviolet stellar occultation measurements of the composition and thermal profiles of the Saturnian upper atmosphere. Geophys Res Lett 9:1147 – 1150

Festou MC, Atreya SK, Donahue TM, Shemansky DE, Sandel BR, Broadfoot AL (1981) Composition and thermal profiles of the Jovian upper atmosphere determined by the Voyager ultraviolet stellar occultation experiment. J Geophys Res 86:5715 – 5725

Fink U, Larson H (1977) The 5 µm spectrum of Saturn. Bull Am Astron Soc 9:535

Fink U, Larson H (1979) The infrared spectra of Uranus, Neptune and Titan from 0.8 to 2.5 microns. Astrophys J 233:1021 – 1040

Fink U, Larson HP, Gautier III TN (1976) New upper limits for atmospheric constituents on Io. Icarus 27:439 – 446

Fjeldbo G, Kliore A, Seidel B, Sweetnam D, Cain D (1975) The Pioneer 10 occultation measurements of the ionosphere of Jupiter. Astron Astrophys 39:91 – 96

Fjeldbo G, Kliore A, Seidel B, Sweetnam D, Woiceshyn P (1976) The Pioneer 11 radio occultation measurements of the Jovian ionosphere. In: Gehrels T (ed) Jupiter. Univ Arizona Press, Tucson, pp 238 – 245

Flasar FM, Samuelson RE, Conrath BJ (1981) Titan's atmosphere: temperature and dynamics. Nature 292:693 – 698

Fowles P, deSorgo M, Yarwood AJ, Strausz OP, Gunning HE (1967) The reactions of sulfur atoms. IX. The flash photolysis of carbonyl sulfide and the reactions of $S(^1D)$ atoms with hydrogen and methane. J Am Chem Soc 89:1352 – 1359

Fox K, Mantz AE, Owen T, Rae KN (1972) A tentative identification of $^{13}CH_4$ and an estimate of $^{12}C/^{13}C$ in the atmosphere of Jupiter. Astrophys J 176:L81 – L84

French RG, Gierasch PJ (1974) Waves in the Jovian upper atmosphere. J Atmos Sci 31:1707 – 1712

French RG, Elliot JL, Dunham EW, Allen DA, Elias JH, Frogel JA, Liller W (1983a) The thermal structure and energy balance of the Uranian upper atmosphere. Icarus 53:399 – 414

French RG, Elias JH, Mink DJ, Elliot JL (1983b) The structure of Neptune's upper atmosphere. Icarus 55:332 – 336

Fricke KH, Darius J (1982) Proc 3rd Eur Conf, ESA SP-176

Frommhold L, Birnbaum G (1984) Hydrogen dimer structures in the F.I.R. spectra of Jupiter and Saturn. Astrophys J 283:L79 – L82

Gardner EP (1981) The vacuum ultraviolet photolysis of methylamine with application to the outer planets and Titan. Thesis, Univ Maryland

Gautier D, Grossman K (1972) A new method for determining the mixing ratio hydrogen to helium in the giant planets. J Atmos Sci 29:788 – 792

Gautier D, Owen T (1983) Elemental and isotopic abundances in the atmospheres of the giant planets: cosmogonical implications. Nature 304:691 – 694

Gautier D, Owen T (1985) Observational constraints on models for giant planet formation. In: Gehrels T (ed) Protostars and Planets. Univ Arizona Press, Tucson

Gautier D, Conrath BJ, Flasar M, Hanel R, Kunde V, Chedin A, Scott N (1981) The helium abundance of Jupiter from Voyager. J Geophys Res 86:8713 – 8720

Gautier D, Bézard B, Marten A, Baluteau JP, Scott N, Chedin A, Kunde V, Hanel R (1982) The C/H ratio in Jupiter from the Voyager infrared investigation. Astrophys J 257:901 – 912

Gautier D, Marten A, Baluteau, Bachet G (1983) About unidentified features in the Voyager far infrared spectra of Jupiter and Saturn. Canad J Phys 61:1455 – 1461

Gehrels N, Stone EC (1983) Energetic oxygen and sulfur ions in the Jovian magnetosphere and their contribution to the auroral excitation. J Geophys Res 88:5537 – 5550

Gehring VM, Hoyermann K, Wagner HG, Wolfrum J (1971) Die Reaktion von atomarem Wasserstoff mit Hydrazin. Ber Bunsenges Phys Chem 73:956

Giauque WF, Blue RW (1936) Hydrogen sulfide. The heat capacity and vapor pressure of solid and liquid. The heat of vaporization. A comparison of thermodynamic and spectroscopic values of the entropy. J Am Chem Soc 58:831 – 837

Giauque WF, Clayton JO (1933) The heat capacity and entropy of nitrogen. Heat of vaporization. Vapor pressures of solid and liquid. The reaction $1/2\,N_2 + 1/2\,O_2 = NO$ from spectroscopic data. J Am Chem Soc 55:4875 – 4889

Gierasch PJ (1983) Dynamical consequences of orthohydrogen-parahydrogen disequilibrium on Jupiter and Saturn. Science 219:847 – 849

Gierasch PJ, Goody RM (1969) Radiative time constants in the atmosphere of Jupiter. J Atmos Sci 26:979 – 980

Gierasch P, Goody RM, Stone P (1970) The energy balance of planetary atmospheres. Geophys Fluid Dyn 1:1 – 18

Giles JW, Moos HW, McKinney WR (1976) The far-ultraviolet (1200 – 1900 Å) spectrum of Jupiter obtained with a rocket-borne multichannel spectrometer. J Geophys Res 81:5797 – 5806

Gillett FC (1975) Further observations of the 8 – 13 micron spectrum of Titan. Astrophys J 201:L41 – L43

Gillett FC, Forrest W (1974) The 7.5 to 13.5 micron spectrum of Saturn. Astrophys J 187:L37 – L38

Gillett FC, Forrest W, Merrill KM (1973) 8 – 13 micron observations of Titan. Astrophys J 184:L93 – L95

Giver LP, Spinrad H (1966) Molecular hydrogen features in the spectra of Saturn and Uranus. Icarus 5:586 – 589

Gladstone GR (1982) Radiative transfer and photochemistry in the upper atmosphere of Jupiter. Thesis, California Inst Technol

Gladstone GR, Yung YL (1983) An analysis of the reflection spectrum of Jupiter from 1500 Å to 1740 Å. Astrophys J 266:415 – 424

Golomb D, Watanabe K, Marmo FF (1962) Absorption coefficients of sulfur dioxide in the vacuum ultraviolet. J Chem Phys 36:958

Goody RM (1964) Atmospheric radiation. Clarendon, Oxford

Gorden R, Ausloos P (1967) Gas phase photolysis and radiolysis of methane: formation of hydrogen and ethylene. J Chem Phys 46:4823 – 4834

Gordon S, Mulac W, Nangia P (1971) Pulse radiolysis of ammonia gas. II. Rate of disappearance of the NH (X^2B_1) radical. J Phys Chem 78:2087 – 2093

Gross SH, Ramanathan GV (1976) The atmosphere of Io. Icarus 29:493 – 502

Gulkis S, De Pater I (1984) A review of the millimeter and centimeter observations of Uranus. In: Bergstralh J (ed) Uranus and Neptune. NASA Conf Publ 2330:225 – 262

Gulkis S, Poynter R (1972) Thermal radio emission from Jupiter and Saturn. Phys Earth Planet Inter 6:36 – 43

Gulkis S, McDonough T, Craft H (1969) The microwave spectrum of Saturn. Icarus 10:421 – 427

Gulkis S, Janssen MA, Olsen ET (1978) Evidence for the detection of ammonia in the Uranus atmosphere. Icarus 34:10 – 19

Gulkis S, Olsen ET, Klein JM, Thompson TJ (1983) Uranus: variability of the microwave spectrum. Science 221:453 – 455

Halstead MP, Leathard DA, Marshall RM, Purnell JH (1970) The reaction of hydrogen atoms with ethylene. Proc R Soc Lond A 316:575 – 591

Ham DI, Trainor DW, Kaufman F (1970) Gas phase kinetics of $H + H + H_2 \rightarrow 2\,H_2$. J Chem Phys 53:4395 – 4396

Hampson RF, McNesby JR (1965) Vacuum ultraviolet photolysis of ethane at high temperature. J Chem Phys 42:2200

Handbook of Chemistry and Physics (1984) CRC, Cleveland, Ohio

Hanel RA, Conrath BJ, Flasar M, Kunde V, Lowman P, Maguire W, Pearl J, Pirraglia J, Samuelson R, Gautier D, Gierasch P, Kumar S, Ponnamperuma C (1979a) Infrared observations of the Jovian system from Voyager 1. Science 204:972 – 976

Hanel RA, Conrath BJ, Flasar M, Herath L, Kunde V, Lowman P, Maguire W, Pearl J, Pirraglia J, Samuelson R, Gautier D, Gierasch P, Horn L, Kumar S, Ponnamperuma C (1979b) Infrared observations of the Jovian system from Voyager 2. Science 206:952 – 956

Hanel RA, Conrath BJ, Flasar FM, Kunde V, Maguire W, Pearl J, Pirraglia J, Samuelson R, Herath L, Allison M, Cruikshank D, Gautier D, Gautier P, Horn L, Koppany R, Ponnamperuma C (1981a) Infrared observations of the Saturnian system from Voyager 1. Science 212:192 – 200

Hanel RA, Conrath BJ, Herath LW, Kunde VG, Pirraglia JA (1981b) Albedo, heat, and energy balance of Jupiter: preliminary results of Voyager infrared investigations. J Geophys Res 86:8705 – 8712

Hanel RA, Conrath BJ, Flaser FM, Kunde V, Maguire W, Pearl J, Pirraglia J, Samuelson R, Herath L, Allison M, Cruikshank D, Gautier D, Gierasch P, Horn L, Koppany R, Ponnamperuma C (1982) Infrared observations of the Saturnian system from Voyager 2. Science 215:544 – 548

Hanel RA, Conrath BJ, Kunde VG, Pearl JC, Pirraglia JA (1983) Albedo and internal heat flux and energy balance of Saturn. Icarus 53:262 – 285

Hanel RA, Conrath BJ, Flasar F, Kunde V, Maguire W, Pearl J, Pirraglia J, Samuelson R, Cruikshank D, Gautier D, Gierasch P, Horn L, Schulte P (1986) Infrared observations of the Uranian system. Science 233:70 – 74

Hanley HJM, McCarty RD, Inteman H (1970) The viscosity and thermal conductivity of dilute gaseous hydrogen from 15 to 5000 K. J Res Nat Bur Stand 74A:331 – 353

Hartle RE, Sittler EC, Ogilvie KW, Scudder JD, Lazarus AJ, Atreya SK (1982) Titan's Ion Exosphere observed from Voyager 1. J Geophys Res 87:1383 – 1394

Hays PB, Roble RG (1968) Atmospheric properties from the inversion of planetary occultation data. Planet Space Sci 16:1197 – 1198

Hays PB, Roble RG, Shah AN (1972) Terrestrial atmospheric composition from stellar occultations. Science 176:793 – 794

Henry RJW, McElroy MB (1968) Photoelectrons in planetary atmospheres. In: Brandt JC, McElroy MB (eds) The atmospheres of Venus and Mars. Gordon and Breach, New York, pp 251 – 285

Henry RJW, McElroy MB (1969) The absorption of extreme ultraviolet solar radiation by Jupiter's upper atmosphere. J Atmos Sci 26:912 – 917

Herzberg G (1966) Molecular spectra and molecular structure. II. Infrared and Raman spectra of polyatomic molecules. Van Nostrand, New York

Herzberg G, Innes KK (1957) Ultraviolet absorption spectra of HCN and DCN. Can J Phys 35:842 – 879

Hildebrand RH, Loewenstein RF, Harper DA, Orton GS, Keene J, Whitcomb SE (1985) Far-infrared and submillimeter brightness temperatures of the giant planets. Icarus 64:64 – 87

Hildendorff HJ (1935) Die Absorptionsspektren von Blausäure, Hydrazin. Äthylen und Ammoniak im Schumann-Gebiet und von Hydrazin im Quarzultraviolett. Z Phys 95:781 – 788

Hinteregger HE (1981) Representations of solar EUV fluxes for aeronomical applications. Adv Space Res 1:39 – 52

Hui MH, Rice SA (1972) Decay of fluorescence from single vibronic states of SO_2. Chem Phys Lett 17:474 – 478

Hunten DM (1969) The upper atmosphere of Jupiter. J Atmos Sci 26:826 – 834

Hunten DM (1971) Composition and structure of planetary atmospheres. Space Sci Rev 12:539 – 599

Hunten DM (1973a) The escape of H_2 from Titan. J Atmos Sci 30:726 – 732

Hunten DM (1973b) The escape of light gases from planetary atmospheres. J Atmos Sci 30:1481 – 1494

Hunten DM (1975) Vertical transport in atmosphere. In: McCormac BM (ed) Atmosphere of earth and planets. Reidel, Dordrecht, Netherlands, pp 59 – 72

Hunten DM (1978) A Titan atmosphere with a surface of 200 K. In: Hunten DM, Morrison D (eds) The Saturn system. NASA Conf Publ 2068, pp 127 – 140

Hunten DM (1984) Atmospheres of Uranus and Neptune. In: Bergstralh T (ed) Uranus and Neptune. NASA Conf Publ 2330, US Government Printing Office, Wash DC

Hunten DM, Dessler AJ (1977) Soft electrons as a possible heat source of Jupiter's thermosphere. Planet Space Sci 25:817 – 821

Hunten DM, Donahue TM (1976) Hydrogen loss from the terrestrial planets. Annu Rev Earth Planet Sci 4:265 – 292

Hunten DM, Veverka J (1976) Stellar and spacecraft occultations by Jupiter: a critical review of derived temperature profiles. In: Gehrels T (ed) Jupiter. Univ Arizona Press, pp 247 – 283

Hunten DM, Watson AJ (1982) Stability of Pluto's atmosphere. Icarus 51:665 – 667

Hunten DM, Tomasko M, Wallace L (1980) Low-latitude thermal structure of Jupiter in the 0.1 – 5 bars. Icarus 43:143 – 152

Hunten DM, Tomasko M, Flasar FM, Samuelson RE, Strobel DF, Stevenson DJ (1984) Titan. In: Gehrels T, Matthews MS (eds) Saturn. Univ Arizona Press, pp 671 – 759

Huntress WT Jr (1974) A review of Jovian ionospheric chemistry. Adv Atmos Mol Phys 10:295 – 340

Huntress WT Jr (1977) Laboratory studies of bimolecular reactions of positive ions in interstellar clouds, in comets, and in planetary atmospheres reducing composition. Astrophys J Suppl 33:495 – 514

Huntress WT, McEwan MJ, Karpas Z, Anicich VG (1980) Laboratory studies of some of the major ion-molecule reactions occurring in cometary comae. Astrophys J Suppl Ser 44:481 – 488

Ingersoll AP, Munch G, Neugebaurer G, Orton GS (1976) Results of the infrared radiometer experiment on Pioneer 10 and 11. In: Gehrels T (ed) Jupiter. Univ Arizona Press, pp 197 – 207

Ingersoll AP, Beebe RF, Conrath BJ, Hunt GE (1984) Structure and dynamics of Saturn's atmosphere. In: Gehrels T, Matthews MS (eds) Saturn. Univ Arizona Press, Tucson, pp 195 – 238

International Critical Tables (1928) McGraw-Hill Book, New York

Ip WH (1983) On plasma transport in the vicinity of the rings of Saturn: a siphon flow mechanism. J Geophys Res 88:819 – 822

Jaffe W, Caldwell J, Owen T (1980) Radius and brightness temperature observations of Titan at centimeter wavelengths by the very large array. Astrophys J 242:806 – 811

Jeans JH (1954) The dynamic theory of gases, 4th edn. Dover, New York

Johnsen R, Biondi MA (1974) Measurements of positive ion conversion and removal reactions relating to the Jovian ionosphere. Icarus 23:139 – 142

Johnson RE, Lanzerotti LJ, Brown WL, Armstrong TP (1981) Erosion of Galilean satellite surfaces by Jovian magnetospheric particles. Science 212:1027 – 1030

Johnson TV, Soderblom LA (1982) Volcanic eruptions on Io: implications for surface evolution and mass loss. In: Morrison D (ed) Satellites of Jupiter. Univ Arizona Press, pp 634 – 646

Johnson TV, Matson DL, Carlson RW (1976) Io's atmosphere and ionosphere: new limits on surface pressure from plasma models. Geophys Res Lett 3:293 – 296

Joyce RR, Pilcher CB, Cruikshank DP, Morrison D (1977) Evidence for weather on Neptune. I. Astrophys J 214:657 – 662

Judge DL, Carlson RW (1974) Pioneer 10 observations of the ultraviolet glow in the vicinity of Jupiter. Science 183:317 – 318

Kaiser ML, Connerney JEP, Desch MD (1983) Atmospheric storm explanation of Saturn electrostatic discharges. Nature 303:50 – 53

Kaiser ML, Desch MD, Connerney JEP (1984) Saturn's ionosphere inferred electron densities. J Geophys Res 89:2371 – 2376

Karpas Z, Anicich VG, Huntress WT Jr (1979) An ion cyclotron resonance study of reactions of ions with hydrogen atoms. J Chem Phys 70:2877 – 2881

Karwat E (1924) Der Dampfdruck des festen Chlorwasserstoffs, Methans und Ammoniaks. Z Phys Chem Stöchiomet Verwandtschaftsr 112:486 – 490

Kaye JA, Strobel DF (1983a) Formation and photochemistry of methylamine in Jupiter's atmosphere. Icarus 55:399 – 419

Kaye JA, Strobel DF (1983b) HCN formation on Jupiter. The coupled photochemistry of ammonia and acetylene. Icarus 54:417 – 433

Keil DG, Lynch KP, Cowfer JA, Michael JV (1976) An investigation of nonequilibrium kinetic isotope effects in chemically activated vinyl radicals. Int J Chem Kinet 8:825 – 857

Kiess CC, Corliss CH, Kiess HK (1960) High dispersion spectra of Jupiter. Astrophys J 132:221 – 231

Kim SJ, Caldwell J, Rivolo AR, Wagener R, Orton GS (1985) Infrared polar brightness of Jupiter 3. Spectrometry from the Voyager 1 IRIS experiment. Icarus 64:233 – 248

Kistiakowski GB, Roberts EK (1953) Rate of association of methyl radicals. J Chem Phys 21:1637 – 1639

Kliore A, Cain DL, Fjeldbo G, Seidel BL (1974) Preliminary results on the atmosphere of Io and Jupiter from the Pioneer 10 s-band occultation experiment. Science 183:323 – 324

Kliore A, Fjeldbo G, Seidel BL, Sweetnam DN, Sesplaukis TT, Woiceshyn PM (1975) Atmosphere of Io from Pioneer 10 radio occultation measurements. Icarus 24:407 – 410

Kliore AJ, Lindal GF, Patel IR, Sweetnam DN, Hotz HB (1980a) Vertical structure of the ionosphere and upper neutral atmosphere of Saturn from the Pioneer radio occultation. Science 207:446 – 449

Kliore AJ, Patel IR, Lindal GF, Sweetnam DN, Hotz HB, Waite JH Jr, McDonough TR (1980b) Structure of the ionosphere and atmosphere of Saturn from Pioneer 11 radio occultation. J Geophys Res 85:5857 – 5870

Knacke RF, Kim SJ, Ridgway ST, Tokunaga AT (1982) The abundances of CH_4, CH_3D, $NH_3 + PH_3$ in the troposphere of Jupiter derived from high-resolution $1100 – 1200$ cm^{-1} spectra. Astrophys J 262:388 – 395

Kostiuk T, Mumma MJ, Espenak F, Deming D, Jennings DE, Maguire W, Zipoy D (1983) Measurements of stratospheric ethane in the Jovian south polar region from infrared heterodyne spectroscopy of the ν_9 band near 12 µm. Astrophys J 265:564 – 569

Kostiuk T, Espenak F, Deming D, Mumma MJ, Zipoy D (1985) Spatial distribution of ethane on Jupiter. Bull Am Astron Soc 17:708

Kuhn WR, Atreya SK, Chang S (1977) The distribution of methylamine in the Jovian atmosphere. Geophys Res Lett 4:203 – 206

Kuiper GP (1944) Titan: a satellite with an atmosphere. Astrophys J 100:378 – 383

Kuiper GP (1952) Planetary atmospheres and their origins. The atmospheres of the Earth and Planets, revised edn. Univ Chicago, pp 306 – 405

Kumar S (1980) A model of the SO_2 atmosphere and ionosphere of Io. Geophys Res Lett 7:9 – 12

Kumar S (1982) Photochemistry of SO_2 in the atmosphere of Io and implications on atmospheric escape. J Geophys Res 87:1677 – 1684

Kumar S, Hunten DM (1982) The atmospheres of Io and other satellites. In: Morrison D (ed) The satellites of Jupiter. Univ Arizona Press, Tucson, pp 782 – 806

Kunde VG, Aikin AC, Hanel RA, Jennings DE, Maguire WC, Samuelson RE (1981) C_4H_2, HC_3N and C_2H_2 in Titan's atmosphere. Nature 292:686 – 688

Kunde VG, Hanel RA, Maguire W, Gautier D, Baluteau JP, Marten AA, Chedin A, Husson N, Scott N (1982) The tropospheric gas composition of Jupiter's north equatorial belt (NH_3, PH_3, CH_3D, GeH_4, H_2O) and the Jovian D/H isotopic ratio. Astrophys J 263:443 – 467

Kupo I, Mekler Y, Eviatar A (1976) Detection of ionized sulfur in the Jovian magnetosphere. Astrophys J 205:L51 – L53

Kurylo MJ, Peterson MC, Braun W (1971) Absolute rate of reaction between H and H_2S. J Chem Phys 54:943 – 945

Lane AL, Hamrick E, Boggess A, Evans DC, Gull TR, Schiffer FH, Turnrose B, Perry P, Holm A, Macchett F (1978) IUE observations of solar system objects. Nature 275:414 – 415

Lane AL, Owen T, Nelson RM, Motteler FC (1979) Ultraviolet spectral variations on Io: an indicator of volcanic activity? Bull Am Astron Soc 11:597

Lanzerotti LJ, Brown WL, Poate JM, Augustyniak WM (1978) On the contribution of water products from Galilean satellites to the Jovian magnetosphere. Geophys Res Lett 5:155 – 158

Larson HP, Fink U, Treffers R, Gautier TN (1975) Detection of water vapor on Jupiter. Astrophys J 197:L137 – L140

Larson HP, Fink U, Treffers R (1976) High resolution (0.5 cm^{-1}) 5-micron spectra of Jupiter from Kuiper airborne observatory. Paper presented at the 7th annual DPS meeting, Austin, Texas, April 1

Larson HP, Treffers RR, Fink U (1977) Phosphine in Jupiter's atmosphere: the evidence from high altitude observations at 5 microns. Astrophys J 211:972 – 979

Larson HP, Fink U, Treffers RR (1978) Evidence for CO in Jupiter's atmosphere from airborne spectroscopic observations at 5 microns. Astrophys J 219:1084 – 1092

Larson HP, Fink U, Smith HA, Davis DS (1980) The middle-infrared spectrum of Saturn: evidence for phosphine and upper limits to other trace atmospheric constituents. Astrophys J 240:327 – 337

Larson HP, Davis DS, Hofmann R, Bjoraker GL (1984) The Jovian atmospheric window at 2.7 μm: a search for H_2S. Icarus 60:621 – 639

Laufer AH (1981a) Kinetics of gas phase reaction of methylene. Rev Chem Intermed 4:225 – 257

Laufer AH (1981b) Reactions of ethynyl radicals. Rate constants with CH_4, C_2H_6, and C_2D_6. J Phys Chem 85:3828 – 3831

Laufer AH (1982) Kinetics and photochemistry of planetary atmospheres. Planetary atmospheres principal investigators meeting, Univ Michigan, Ann Arbor, Michigan (abstract)

Lecacheux J, deBergh C, Combes M, Maillard J (1976) The C/H and $^{12}CH_4/^{13}CH_4$ in the atmospheres of Jupiter and Saturn from 0.1 cm resolution near-infrared spectra. Astron Astrophys 53:29 – 33

Lee JH, Michaels JV, Payne WA, Stief LJ (1978) Absolute rate of the reaction of atomic hydrogen with ethylene from 198 to 320 K at high pressure. J Chem Phys 68:1817 – 1820

Lee LC (1980) CN($A^2\pi - X^2\sigma$) and CN($B^2\sigma - X^2\sigma$) yields from HCN photodissociation. J Chem Phys 72:6414 – 6421

Leovy CB, Pollack JB (1973) A first look at atmospheric dynamics and temperature variations on Titan. Icarus 19:195 – 201

Leu MT, Biondi MA, Johnsen R (1973) Measurements of the recombination of electrons with H_3^+ and H_5^+ ion. Phys Rev 8:413 – 419

Lewis JS (1969) The clouds of Jupiter and the $NH_3 - H_2O$ and $NH_3 - H_2S$ systems. Icarus 10:365 – 378

Lewis JS (1972) Low temperature condensation from the solar nebula. Icarus 16:241 – 252

Lewis JS (1980) Lightning synthesis of organic compounds on Jupiter. Icarus 43:85 – 95

Lewis JS, Prinn RG (1970) Jupiter's clouds: structure and composition. Science 169:472 – 473

Lewis JS, Prinn RG (1980) Kinetic inhibition of CO and N_2 reduction in the solar nebula. Astrophys J 238:357 – 364

Lias SG, Collin GJ, Rebbert RE, Ausloos P (1970) Photolysis of ethane at 11.6 – 11.8 eV. J Chem Phys 52:1213 – 1219

Lindal GF, Wood GE, Levy GS, Anderson JD, Sweetnam DN, Hotz HB, Buckles BJ, Holmes DF, Doms PE, Eshleman VR, Tyler GL, Croft TA (1981) The atmosphere of Jupiter: an analysis of the Voyager radio occultation measurements. J Geophys Res 86:8721 – 8727

Lindal GF, Wood GE, Hotz HB, Sweetnam DM, Eshleman VR, Tyler GL (1983) The atmosphere of Titan: an analysis of the Voyager 1 ratio occultation measurements. Icarus 53:348 – 363

Lindal GF, Sweetnam DN, Eshleman VR (1985) The atmosphere of Saturn: an analysis of the Voyager radio occultation measurements. Astron J 90:1136 – 1146

Lindzen RS (1971) Tides and gravity in the upper atmosphere. In: Fiocco G (ed) Mesospheric models and related experiments. Reidel, Dordrecht, Holland, pp 122 – 130

Linke WF (1965) Solubilities of inorganic and metalorganic compounds, 4th edn. Am Chem Soc, Wash DC

Liu SC, Donahue TM (1975) The aeronomy of the upper atmosphere of Venus. Icarus 24:148 – 156

Lunine JI, Stevenson DJ (1982) Post-accretional evolution of Titan's surface and atmosphere. Astrophys J 238:357 – 364

Lunine JI, Stevenson DJ, Yung YL (1983) Ethane ocean on Titan. Science 222:1229 – 1230

Lutz BL, Owen T (1974) The search for HD in the spectrum of Uranus: an upper limit to [D/H]. Astrophys J 190:731 – 734

Lutz BL, Owen T, Cess RD (1976) Laboratory band strengths of methane and their applications to the atmospheres of Jupiter, Saturn, Uranus, Neptune and Titan. Astrophys J 203: 541 – 551

Lutz BL, DeBergh C, Owen T (1983) Titan: discovery of carbon monoxide in its atmosphere. Science 220:1374 – 1375

Macy WW Jr, Sinton WM (1977) Detection of methane and ethane emission on Neptune but not on Uranus. Astrophys J 218:L79 – L81

Macy WW Jr, Smith WM (1978) Detection of HD on Saturn and Uranus, and the D/H ratio. Astrophys J 222:L73 – L75

Maguire WC, Hanel RA, Jennings DE, Kunde VG, Samuelson RE (1981) C_3H_8 and C_3H_4 in Titan's atmosphere. Nature 292:683 – 686

Maier HN, Fessenden RW (1975) Electron ion recombination rate constants for some compounds of moderate complexity. J Chem Phys 162:4790 – 4795

Martin TZ (1975) Saturn and Jupiter: a study of atmospheric constituents. Thesis, Univ Hawaii, Honolulu

Mason EA, Marrero TR (1970) The diffusion of atoms and molecules. Adv Atmos Mol Phys 6:156 – 232

Massie ST, Hunten DM (1982) Conversion of para and ortho hydrogen in the Jovian planets. Icarus 49:213 – 216

Matson DL, Nash D (1983) Io's atmosphere: pressure control by Regolith cold trapping and surface venting. J Geophys Res 88:4771 – 4783

McConnell JC, Sandel BR, Broadfoot AL (1981) Voyager UV spectrometer observations of He 584 Å dayglow at Jupiter. Planet Space Sci 29:283 – 292

McConnell JC, Holberg JB, Smith GR, Sandel B, Shemansky D, Broadfoot AL (1982) A new look at the ionosphere of Jupiter in light of the UVS occultation results. Planet Space Sci 30:151 – 167

McCord TB (1966) Dynamical evolution of the Neptunian system. Astron J 71:585 – 590

McElroy MB (1973) The ionospheres of the major planets. Space Sci Rev 14:460 – 473

McElroy MB, Yung YL (1975) The atmosphere and ionosphere of Io. Astrophys J 196:227 – 250

McGuire EJ (1968) Photoionization cross-sections of the elements helium to xenon. Phys Rev 175:20 – 24

McNesby JR, Okabe H (1964) Vacuum ultraviolet photochemistry. Adv Photochem 3:157 – 240

McNesby JR, Tanaka I, Okabe H (1962) Vacuum ultraviolet photochemistry. III. Primary processes in the vacuum ultraviolet photolysis of water and ammonia. J Chem Phys 36:605 – 607

McNutt RL (1983) Vector velocities in Saturn's magnetosphere. Paper presented at 5th Conf Phys Jovian Saturnian Magnetosphere. MIT, Cambridge

McNutt RL, Belcher JW, Sullivan JD, Bagenal F, Bridge HS (1979) Departure from rigid corotation of plasma in Jupiter's dayside magnetosphere. Nature 280:803

Michael JV, Osborne DT, Suess GN (1973) Reaction $H + C_2H_4$: investigation into the effects of pressure, stoichiometry, and the nature of third body species. J Chem Phys 58:2800 – 2812

Miller SL (1961) The occurrence of gas hydrates in the solar system. Proc Natl Acad Sci USA 47:1798 – 1808

Miller SL (1973) In: Whalley E, Jones SJ, Gold LW (eds) Physics and chemistry of ice. Univ Toronto Press, R Soc Can, pp 42 – 50

Miller TM, Moseley JT, Martin DW, McDaniel EW (1968) Reactions of H^+ in H_2 and D^+ in D_2: mobilities of hydrogen and alkali ions in H_2 and D_2 gases. Phys Rev 173:115 – 123

Mizuno H (1980) Formation of the giant planets. Prog Theor Phys 64:544 – 557

Mizuno H, Nakazawa K, Hayashi C (1978) Instability of a gaseous envelope surrounding a planetary core and formation of the giant planets. Prog Theor Phys 60:699 – 710

Moos HW (1981) Ultraviolet emissions from the upper atmospheres of the planets. In: Atreya SK, Caldwell JJ (eds) Planetary aeronomy and astronomy. Pergamon, Oxford, pp 155 – 164

Moos HW, Clarke JT (1981) Ultraviolet observations of the Io torus from the IUE observatory. Astrophys J 247:354 – 361

Moos HW, Fastie WG, Bottema M (1969) Rocket measurement of ultraviolet spectra of Venus and Jupiter between 1200 Å and 1800 Å. Astrophys J 155:887 – 897

Morabito LA, Synnott SP, Kupferman PM, Collins SA (1979) Discovery of currently-active extraterrestrial volcanism. Science 204:972

Morfill GE, Fechtig H, Grün E, Goertz CK (1983) Some consequences of meteoroid impacts on Saturn's rings. Icarus 55:439 – 447

Morgan JS, Pilcher CB (1981) Plasma characteristics of the Io torus. Astrophys J 253:406 – 421

Morrison D (1982) Introduction to the satellites of Jupiter. In: Morrison D (ed) Satellites of Jupiter. Univ Arizona Press, pp 3 – 43

Morrison D, Cruikshank DP, Brown RH (1982) Diameters of Triton and Pluto. Nature 300:425 – 427

Moseley SH, Conrath BJ, Silverberg RF (1985) Atmospheric temperature profiles of Uranus and Neptune. Astrophys J 292:L83 – L87

Mount GH, Moos HW (1978) Photoabsorption cross sections of methane and ethane 1380 – 1600 Å at $T = 295$ and 200 K. Astrophys J 224:L35 – L38

Mount GH, Warden ES, Moos HW (1977) Photoabsorption cross sections of methane from 1400 – 1850 Å. Astrophys J 214:L47 – L49

Münch R, Spinrad H (1963) On the spectrum of Saturn. Mem Soc R Liege 7:541 – 542

Munson MSB, Field FH (1969) Reactions of gaseous ions. XVII. Methane and unsaturated hydrocarbons. J Am Chem Soc 91:3413 – 3418

Nagy AF, Chameides WL, Chen RH, Atreya SK (1976) Electron temperatures in the Jovian ionosphere. J Geophys Res 81:5567 – 5569

Nakayama T, Watanabe K (1964) Absorption and photoionization coefficients of acetylene, propyne and l-butyne. J Chem Phys 40:588 – 561

Ness NF, Acuna MH, Lepping RP, Connerney JEP, Behannon KW, Burlaga LF, Neubauer FM (1981) Magnetic field studies by Voyager 1: preliminary results at Saturn. Science 212:211 – 217

Nier AO, McElroy MB (1977) Composition and structure of Mars upper atmosphere: results from the neutral mass spectrometers on Viking 1 and 2. J Geophys Res 82:4341 – 4349

Nishida A, Watanabe Y (1981) Joule heating of the Jovian ionosphere by corotation enforcement currents. J Geophys Res 86:9945 – 9952

Noll K (1985) Detection of CO at Saturn. Paper presented at the Conf Jovian Atmos. Goddard Inst Space Stud, New York, 6 – 8 May

Okabe H (1981) Photochemistry of acetylene at 1470 Å. J Chem Phys 75:2772 – 2778

Okabe H (1983) Photochemistry of acetylene at 1849 Å. J Chem Phys 78:1312 – 1317

Olsen ET, Gulkis S (1978) A preliminary investigation of Neptune's atmosphere via its microwave continuum emission. Bull Am Astron Soc 10:577

Orient OJ, Srivastava SK (1984) Mass spectrometer of partial and total electron impact ionization cross sections of SO_2 from threshold up to 200 eV. J Chem Phys 80:140 – 143

Orton GS, Appleby JF (1984) Temperature structure and infrared derived properties of the atmospheres of Uranus and Neptune. In: Bergstralh TT (ed) Uranus and Neptune. NASA Conf Publ 2330:89 – 156. US Government Printing Office, Wash DC

Orton GS, Ingersoll AP (1976) Pioneer 10 and 11 and ground-based infrared data on Jupiter: the thermal structure and $He – H_2$ ratio. In: Gehrels T (ed) Jupiter. Univ Arizona Press, Tucson, pp 206 – 215

Orton GS, Tokunaga AT, Caldwell JJ (1983) Observational constraints on the atmospheres of Uranus and Neptune from new measurements near 10 microns. Icarus 56:147 – 164

Orton GS, Griffin MJ, Ade PAR, Nolt IG, Radostitz JV, Robson EI, Gear WK (1986) Submillimeter and millimeter observations of Uranus and Neptune. Icarus 67:289 – 304

Owen T (1965) Comparisons of laboratory and planetary spectra. II. The spectrum of Jupiter from 9700 to 11,200 Å. Astrophys J 141:444 – 456

Owen T (1969) The spectra of Jupiter and Saturn in the photographic infrared. Icarus 10:355 – 364

Owen T (1982) The composition and origin of Titan's atmosphere. Planet Space Sci 30:833 – 838

Owen T, McKellar A, Encrenaz T, Lecacheux J, deBergh C, Maillard J (1977) A study of the 1.56 micron NH_3 band on Jupiter and Saturn. Astron Astrophys 54:291 – 295

Owen T, Caldwell JJ, Rivolo AR, Moore V, Lane AL, Sagan C, Hunten G, Ponnamperuma C (1980) Observations of the spectrum of Jupiter from 1500 to 2000 Å with the IUE. Astrophys J 236:L39 – L42

Patrick R, Pilling MJ, Rogers GJ (1980) A high pressure rate constant for $CH_3 + H$ and an analysis of the kinetics of the $CH_3 + H \rightarrow CH_4$ reaction. Chem Phys 53:279 – 291

Payne WA, Stief LJ (1976) Absolute rate constant for the reaction of atomic hydrogen with acetylene over an extended pressure and temperature range. J Chem Phys 64:1150 – 1155

Peale SJ, Cassen P, Reynolds RT (1979) Melting of Io by tidal dissipation. Science 203:892 – 894

Pearl J, Hanel R, Kunde V, Maguire W, Fox K, Gupta S, Ponnamperuma C, Raulin F (1979) Identifications of gaseous SO_2 and new upper limits for other gases on Io. Nature 280:755 – 758

Pearl JC, Sinton WM (1982) Hot spots on Io. In: Morrison D (ed) Satellites of Jupiter. Univ Arizona Press, pp 724 – 755

Phillips LF (1981) Absolute absorption cross-section for SO between 190 and 235 nm. J Phys Chem 85:3994 – 4000

Pilcher CB, Morgan JS (1979) Detection of singly ionized oxygen around Jupiter. Science 205:297 – 298

Pollack JB (1973) Greenhouse models of the atmosphere of Titan. Icarus 19:43 – 58

Pollack JB, Wilteborn F, Erickson E, Streker F, Baldwin B, Bunch T (1978) Near infrared spectra of the Galilean satellites: observations and composition implications. Icarus 36:271 – 303

Pollack JB, Rages K, Bergstralh J, Wenkert D, Danielson G (1985) Vertical structure of aerosols and clouds in the atmosphere of Uranus and Neptune. Paper presented at the IAMAP Scientific Assembly, Honolulu, Hawaii, 12 August

Pollack JB, Rages K, Baines KH, Bergstralh JT, Wenkert D, Danielson GE (1986) Estimates of the bolometric albedoes and radiation balance of Uranus and Neptune. Icarus 65:442 – 466

Prasad SS, Tan A (1974) The Jovian ionosphere. Geophys Res Lett 1:337 – 340

Prather MJ, Logan JA, McElroy MB (1978) Carbon monoxide in Jupiter's upper atmosphere: an extraplanetary source. Astrophys J 223:1072 – 1081

Prinn RG, Barshay SS (1977) Carbon monoxide on Jupiter and implications for atmospheric convection. Science 198:1031 – 1034

Prinn RG, Fegley B Jr (1981) Kinetic inhibition of CO and N reduction in circumplanetary nebulae: implications for satellite composition. Astrophys J 249:308 – 317

Prinn RG, Lewis JS (1975) Phosphine on Jupiter and implications for the Great Red Spot. Science 190:274 – 276

Prinn RG, Olaguer EP (1981) Nitrogen on Jupiter: a deep atmospheric source. J Geophys Res 86:9895 – 9899

Prinn RG, Owen TC (1976) Chemistry and spectroscopy of the Jovian atmosphere. In: Gehrels T (ed) Jupiter. Univ Arizona Press, Tucson, pp 319 – 371

Prinn RG, Larson HP, Caldwell JJ, Gautier D (1984) Composition and chemistry of Saturn's atmosphere. In: Gehrels T, Matthews MS (eds) Saturn. Univ Arizona Press, Tucson, pp 88 – 149

Rebbert RE, Ausloos P (1972) Photolysis of methane quantum yield of $C(^1D)$ and CH. J Photochem 1:171 – 176

Rees MH, Roble RG (1975) Observations and theory of the formation of stable auroral red arcs. Rev Geophys Space Phys 13:201 – 242

Ridgway ST (1974) The infrared spectrum of Jupiter, 750 – 1200 cm. Bull Am Astron Soc 6:376

Ridgway ST, Larson HP, Fink U (1976) The infrared spectrum of Jupiter. In: Gehrels T (ed) Univ Arizona Press, Tucson, pp 384 – 417

Rieke GH, Lebofsky LA, Lebofsky MJ (1985) A search for nitrogen on Triton. Icarus 64:153 – 155

Romani PN, Atreya SK (1985) Clouds in the atmosphere of Uranus. Bull Am Astron Soc 17: 744

Rossow WB (1978) Cloud microphysics: analysis of the clouds of Earth, Venus, Mars and Jupiter. Icarus 36:1 – 50

Rottman GJ, Moos HW, Freer CS (1973) The far-ultraviolet spectrum of Jupiter. Astrophys J 184:L89 – L92

Sagan C, Khare BN (1979) Tholins: organic chemistry of interstellar grains and gas. Nature 227:102 – 107

Sagan C, Khare BN, Lewis JS (1984) Organic matter in the Saturn system. In: Gehrels T (ed) Univ Arizona Press, Tucson, pp 788 – 807

Samuelson RE, Hanel RA, Kunde VG, Maguire WC (1981) Mean molecular weight and hydrogen abundance of Titan's atmosphere. Nature 292:688 – 693

Samuelson RE, Maguire WC, Hanel RA, Kunde VG, Jennings DE, Yung YL, Aikin A (1983) CO_2 on Titan. Icarus 88:8709 – 8715

Sandel BR, Shemansky DE, Broadfoot AL, Bertaux JL, Blamont JE, Belton M, Ajello J, Holberg J, Atreya S, Donahue T, Moos H, Strobel D, McConnell M, Dalgarno A, Goody R, McElroy M, Takacs P (1979) Extreme ultraviolet observations from Voyager 2 encounter with Jupiter. Science 206:962 – 966

Sandel BR, Broadfoot AL, Strobel DF (1980) Discovery of a longitudinal asymmetry in the hydrogen Lyman-alpha brightness of Jupiter. Geophys Res Lett 7:5 – 8

Sandel BR, Shemansky DE, Broadfoot AL, Holberg JB, Smith GR, McConnell JC, Strobel DF, Atreya SK, Donahue TM, Moos HW, Hunten DM, Pomphrey RB, Linick S (1982a) Extreme ultraviolet observations from the Voyager 2 encounter with Saturn. Science 215:548 – 553

Sandel BR, McConnell JC, Strobel DF (1982b) Eddy diffusion at Saturn's homopause. Geophys Res Lett 9:1077 – 1080

Schacke H, Wagner HG, Wolfrum J (1977) Reaktionen von Molekülen in definierten Schwingungszuständen. IV. Reaktionierte schwingungsangeregte Cyan-Radikale mit Wasserstoff und einfachen Kohlenwasserstoffen. Ber Bunsenges Phys Chem 81:670 – 675

Schurath U, Tiedemann P, Schindler R (1969) Photolysis of ammonia at 2062 Å in presence of ethylene. J Phys Chem 73:456 – 469

Science (1979) Mission to Jupiter and its satellites. 204:945 – 1008

Science (1986) Voyager encounter with the Uranus system. 233:39 – 109

Shemansky DE, Smith GR (1981) The Voyager 1 EUV spectrum of the plasma torus. J Geophys Res 86:9176 – 9192

Shemansky DE, Smith GR (1986) The implication for the presence of a magnetosphere on Uranus in the relationship of EUV and radio emission. Geophys Res Lett 13:2 – 5

Shimizu M (1980) Strong interaction between the ring system and the ionosphere of Saturn. Proc 13th Lunar and Planetary Symp Inst Space Aeron Sci, Univ of Tokyo, July 7 – 9

Sicardy B, Combes M, Brahic A, Bouchet P, Perrier C, Courtin R (1982) The 15 August 1980 occultation by the Uranian system: structure of the rings and temperature of the upper atmosphere. Icarus 52:454 – 472

Sicardy B, Combes M, Lecacheux J, Bouchet P, Brahic A, Laques P, Perrier C, Vapillon L, Zeau Y (1985) Variations of the stratospheric temperature profile along the limb of Uranus: results of the 22 April 1982 stellar occultation. Icarus 64:88 – 106

Singer SF, Maeda K (1961) Energy dissipation of spiraling particles in the polar atmospheres. Arkh Geophys 3:531 – 538

Siscoe GL, Chen CK (1977) Io: a source for Jupiter's inner plasmasphere. Icarus 31:1 – 10

Skinner TE (1984) Temporal and spatial variations in the intensity of ultraviolet emissions from Jupiter and the Io torus. Thesis, John Hopkins Univ, Baltimore, Maryland

Slanger TG (1982) 1216 Å photodissociation of H_2O, NH_3 and CH_4. 4th Meet Planet Atmos Princ Investig, Ann Arbor, 21 – 23 April, pp 129 – 131 (abstracts)

Smith BA (1984) Near infrared imaging of Uranus and Neptune. In: Bergstralh J (ed) Uranus and Neptune. NASA Conf Publ 2330:213 – 262

Smith BA, Soderblom LA, Johnson TV, Ingersoll AP, Collins SA, Shoemaker EM, Hunt GE, Masursky H, Carr MH, Davies ME, Cook AF II, Boyce J, Danielson GE, Owen T, Sagan C, Beebe RJ, Veverka J, Strom RG, McGauley JF, Morrison D, Briggs GA, Suomi UE (1979a) The Jupiter system through the eyes of Voyager 1. Science 204:951 – 972

Smith BA, Soderblom L, Beebe R, Boyce J, Briggs G, Carr M, Collins S, Cook A, Danielson G, Davies M, Hunt G, Ingersoll A, Johnson T, Masursky H, McCauley J, Morrison D, Owen T, Sagan C, Shoemaker E, Strom R, Suomi V, Veverka J (1979b) The Galilean satellites and Jupiter: Voyager 2 imaging science results. Science 206:927 – 950

Smith BA, Soderblom L, Beebe R, Boyce J, Briggs G, Bunker A, Collins SA, Hansen CJ, Johnson TV, Mitchell JL, Terrile RJ, Carr M, Cook AFII, Cuzzi J, Pollack JB, Danielson GE, Ingersoll A, Davies ME, Hunt GE, Masursky H, Shoemaker E, Morrison D, Owen T, Sagan C, Veverka J, Strom R, Suomi VE (1981) Encounter with Saturn: Voyager 1 imaging science results. Science 212:163 – 190

Smith BA, Soderblom L, Beebe R, Bliss D, Boyce J, Brahic A, Briggs G, Brown R, Collins S, Cook A II, Croft S, Cuzzi J, Danielson G, Davies M, Dowling T, Godfrey D, Hansen C, Harris C, Hunt G, Ingersoll A, Johnson T, Krauss R, Masursky H, Morrison D, Owen T, Plescia J, Pollack J, Porco C, Rages K, Sagan C, Shoemaker E, Sromovsky L, Stoker C, Strom R, Suomi V, Synnott S, Terrile R, Thomas P, Thompson W, Veverka J (1986) Voyager 2 in the Uranus system: imaging science results. Science 233:43 – 64

Smith DW, Adams NG (1977) Reactions of simple hydrocarbon ions with molecules at thermal energies. Int J Mass Spectrom Ion Phys 23:123 – 135

Smith GR, Strobel DF, Broadfoot AL, Sandel BR, Shemansky DE, Holberg JB (1982) Titan's upper atmosphere: composition and temperature from the EUV solar occultation results. J Geophys Res 87:1351 – 1357

Smith GR, Shemansky DE, Holberg JB, Broadfoot AL, Sandel BR, McConnell JC (1983) Saturn's upper atmosphere from the Voyager 2 EUV solar and stellar occultation. J Geophys Res 88:8667 – 8678

Smith WH (1978) On the ortho-para equilibration of H_2 in the atmospheres of the Jovian planets. Icarus 33:210 – 216

Smoluchowski R (1967) Internal structure and energy emission of Jupiter. Nature 215:691 – 695

Soifer BJ, Neugebauer G, Matthews K (1980) The 1.5 – 2.5 micron spectrum of Pluto. Astron J 85:166 – 167

Spinrad H, Trafton L (1963) High dispersion spectra of the outer planets. I. Jupiter in the visual and red. Icarus 2:19 – 28

Spitzer L Jr (1962) Physics of fully ionized gases. Wiley, Interscience, New York

Stevenson DJ (1980) Saturn's luminosity and magnetism. Science 208:746 – 748

Stevenson DJ, Salpeter EE (1977) The dynamics and helium distribution properties for hydrogen-helium fluid planets. Astrophys J Suppl Ser 35:239 – 261

Stief LJ, Payne WA (1976) Absolute rate parameter for the reaction of hydrogen with hydrazine. J Chem Phys 64:4892 – 4897

Stone PH (1973) The dynamics of the atmospheres of the major planets. Space Sci Rev 14:444 – 459

Strobel DF (1969) The photochemistry of methane in the Jovian atmosphere. J Atmos Sci 26:909 – 911

Strobel DF (1973a) The photochemistry of NH_3 in the Jovian atmosphere. J Atmos Sci 30:1205 – 1209

Strobel DF (1973b) The photochemistry of hydrocarbons in the Jovian atmosphere. J Atmos Sci 30:489 – 498

Strobel DF (1975) Aeronomy of the major planets. Photochemistry of ammonia and hydrocarbons. Rev Geophys Space Phys 13:372 – 382

Strobel DF (1977) NH_3 and PH_3 photochemistry in the Jovian atmosphere. Astrophys J 214:L97 – L99

Strobel DF, Atreya SK (1983) Ionosphere. In: Dessler AJ (ed) Physics of the Jovian magneto-sphere. Cambridge Univ Press, pp 51 – 67

Strobel DF, Shemansky DE (1982) EUV emission from Titan's upper atmosphere: Voyager 1 en-counter. J Geophys Res 87:1361 – 1368

Strobel DF, Smith GR (1973) On the temperature of the Jovian thermosphere. J Atmos Sci 39:718 – 725

Strobel DF, Yung YL (1979) The Galilean satellites as a source of CO in the Jovian upper atmo-sphere. Icarus 37:256 – 263

Strom RG, Schneider NM (1982) Volcanic eruption plumes on Io. In: Morrison D (ed) Satellites of Jupiter. Univ Arizona Press, Tucson, pp 598 – 633

Stubbe P (1968) Frictional forces and collision frequencies between moving ion and neutral gases. J Atmos Terr Phys 30:1965 – 1968

Stull DR (1947) Vapor pressure of pure substance: organic compounds. Indust Engin Chem 39:517 – 540

Teng L, Jones WE (1972) Kinetics of the reactions of hydrogen atoms with ethylene and vinyl flu-oride. J Chem Soc Faraday Trans 68:1267

Theard LP, Huntress WT Jr (1974) Ion-molecule reactions and vibrational deactivation of H_2^+ ions in mixtures of hydrogen and helium. J Chem Phys 60:2840 – 2848

Thompson BA, Harteck P, Reeves R Jr (1963) Ultraviolet absorption coefficients of CO_2, CO, O_2, H_2O, N_2O, NH_3, NO, SO_2 and CH_4 between 1850 and 4000 Å. J Geophys Res 68:6431 – 6436

Tokunaga AT, Knacke R, Owen T (1975) The detection of ethane on Saturn. Astrophys J 197:L77 – L78

Tokunaga AT, Beck SC, Geballe TR, Lacy JH, Serabyn E (1981) The detection of HCN on Jupiter. Icarus 48:283 – 289

Tokunaga AT, Orton GS, Caldwell J (1983) New observational constraints on the temperature inversions of Uranus and Neptune. Icarus 53:141 – 146

Trafton LM (1967) Model atmospheres of the major planets. Astrophys J 147:765 – 781

Trafton LM (1972) The bulk composition of Titan's atmosphere. Astrophys J 175:295 – 306

Trafton LM (1973) Saturn: as study of the 3 μm methane band. Astrophys J 182:615 – 636

Trafton LM (1974) The source of Neptune's internal heat and the value of Neptune's tidal dissi-pation factor. Astrophys J 193:477 – 480

Trafton LM (1975) Detection of a potassium cloud near Io. Nature 258:690 – 692

Trafton LM (1976) The aerosol distribution in Uranus atmosphere: interpretation of the hydro-gen spectrum. Astrophys J 207:1007 – 1024

Trafton LM (1977) Periodic variations in Io's sodium and potassium clouds. Astrophys J 215:960 – 970

Trafton LM (1978) Detection of H_2 quadrupole lines belonging to the (5 – 0) overtone band in the spectrum of Uranus. Astrophys J 223:339 – 343

Trafton LM (1981) The atmospheres of the outer planets and satellites. Rev Geophys Space Phys 19:43 – 89

Trafton LM, Macy W (1975) Saturn's 3 v_3 methane band: an analysis in terms of a scattering at-mosphere. Astrophys J 196:867 – 876

Trafton LM, Ramsay DA (1980) The D/H ratio in the atmosphere of Uranus: detection of the $R_5(1)$ line of HD. Icarus 41:423 – 429

Trauger JT, Roesler FL, Michelson M (1977) The D/H ratios on Jupiter, Saturn and Uranus based on new HD and H_2 data. Bull Am Astron Soc 9:516

Trauger JT, Munch G, Roesler FL (1979) A study of the Jovian [SII] and [SIII] nebulae at high spectral resolution. Bull Am Astron Soc 11:591 – 592

Troe J (1977) Theory of thermal unimolecular reactions at low pressures. II. Strong collisions rate constants, applications. J Chem Phys 66:4758 – 4775

Tyler GL, Eshleman VR, Anderson JD, Levy GS, Lindal GF, Wood GE, Croft TA (1981) Radio science investigation of the Saturn system with Voyager 1: preliminary results. Science 212:201 – 206

Tyler GL, Eshleman VR, Anderson JD, Levy GS, Lindal GF, Wood GE, Croft TA (1982) Radio science with Voyager 2 at Saturn: atmosphere and ionosphere and the mass of Mimas, Tethys and Iapetus. Science 215:553–557

Tyler GL, Sweetnam DN, Anderson JD, Campbell JK, Eshleman VR, Hinson DP, Levy GS, Lindal GF, Marouf EA, Simpson RA (1986) Radio science observations of the Uranian system with Voyager 2: properties of the atmosphere, rings and satellites. Science 233:79–84

van den Bergh HE (1976) The recombination of methyl radicals in the low pressure limit. Chem Phys Lett 43:201–204

Vera Ruiz HG, Rowland FS (1978) Possible scavenging reactions of C_2H_2 and C_2H_4 for phosphorus-containing radicals in the Jovian atmosphere. Geophys Res Lett 5:407–410

Veverka JL, Wasserman LH, Elliot J, Sagan C, Liller W (1974) The occultation of β Scorpii by Jupiter. I. The structure of the Jovian atmosphere. Astron J 79:73–84

Von Homann KH, Krome G, Wagner HG (1968) Schwefelkohlenstoff-Oxidation, Geschwindigkeit von Elementarreaktionen. I. Ber Bunsenges Phys Chem 22:998

von Zahn U, Fricke KH, Hunter DM, Kraukowsky D, Mauersberger K, Nier AO (1980) The upper atmosphere of Venus during morning conditions. J Geophys Res 85:7829–7840

Wagman DD (1979) Data sheet. Chem Thermodynam Data Center, Natl Bur Stand Wash DC

Waite JH Jr (1981) The ionosphere of Saturn. PhD Thesis, Univ Michigan

Waite JH Jr, Cravens TE, Kozyra JU, Nagy AF, Atreya SK, Chen RH (1983) Electron precipitation and related aeronomy of the Jovian thermosphere and ionosphere. J Geophys Res 88:6143–6163

Wallace L (1976) The thermal structure of Jupiter in the stratosphere and upper troposphere. In: Gehrels T (ed) Jupiter. Univ Arizona Press, pp 284–303

Wallace L (1980) The structure of the Uranus atmosphere. Icarus 43:231–259

Wallace L, Hunten DM (1973) The Lyman-alpha albedo of Jupiter. Astrophys J 182:1013–1031

Warwick JW, Pearce JB, Riddle AC, Alexander JK, Desch MD, Kaiser ML, Thieman JR, Carr TD, Bulkis S, Boischot A, Harvey CC, Pedersen BM (1979) Voyager 1 planetary radio astronomy observations near Jupiter. Science 204:995–998

Warwick JW, Pearce J, Evans D, Carr T, Schauble J, Alexander J, Kaiser M, Desch M, Pedersen M, Lecacheux A, Daigne G, Boischot A, Barrow CH (1981) Planetary radio astronomy observations from Voyager 1 near Saturn. Science 212:239–243

Washburn EW (1924) The vapor pressure of ice and of water below the freezing point. Monthly Weather Review, pp 488–490

Watanabe K, Zelikoff M, Lin ECY (1953) Absorption coefficients of several atmospheric gases. Tech Rep 53:23–79

Weidenschilling SJ, Lewis JS (1973) Atmospheric and cloud structure of the Jovian planets. Icarus 20:465–476

Weiser H, Vitz FRC, Moos HW (1977) Detection of Lyman-alpha emission from the Saturnian disc and from the ring system. Science 197:755–757

West RA, Hord CW, Simmons KE, Hart H, Esposito LW, Lane AL, Pomphrey RB, Morris RB, Sato M, Coffeen D (1983) Voyager photopolarimeter observations of Saturn and Titan. Adv Space Res 3:45–48

Whitten RC, Reynolds TR, Michelson PF (1975) The ionosphere and atmosphere of Io. Geophys Res Lett 2:49–51

Wildt R (1932) Absorptionsspektren und Atmosphären der grossen Planeten. Veröff Univ Sternwarte Göttingen 22:171–180

Wildt R (1937) Photochemistry of planetary atmospheres. Astrophys J 86:321–336

Wilson TA (1925) The total and partial vapor pressures of aqueous ammonia solutions. Univ Illinois Engin Exp Station. Bulletin No. 146

Winkelstein P, Caldwell J, Owen T, Combes M, Encrenaz T, Hunt G, Moore V (1983) A determination of the composition of the Saturnian stratosphere using IUE. Icarus 54:309–318

Woeller F, Ponnamperuma C (1969) Organic synthesis in a simulated Jovian atmosphere. Icarus 10:386–392

Woodman JH, Trafton L, Owen T (1977) The abundance of ammonia in the atmospheres of Jupiter, Saturn, and Titan. Icarus 32:314–320

Wu R, Judge DL (1981) Study of sulfur containing molecules in the EUV region. I. Photoabsorption cross-section of SO_2. J Chem Phys 74:3804 – 3806

Young AT (1984) No sulfur flows on Io. Icarus 58:197 – 226

Yung YL, McElroy MB (1977) Stability of an oxygen atmosphere on Ganymede. Icarus 30:97 – 103

Yung YL, Strobel DF (1980) Hydrocarbon photochemistry and Lyman-alpha albedo of Jupiter. Astrophys J 239:395 – 402

Yung YL, Allen M, Pinto JP (1984) Photochemistry of the atmosphere of Titan: comparison between model and observations. Astrophys J Suppl 55:465 – 506

Zahnle KJ, Walker JCG (1982) The evolution of solar luminosity. Rev Geophys Space Phys 20:280 – 292

Zelikoff M, Watanabe K (1953) Absorption coefficients of ethylene in the vacuum ultraviolet. J Opt Soc Am 43:756 – 759

Ziegler WT (1959) The vapor pressures of some hydrocarbons in the liquid and solid state at low temperatures. Natl Bur Stand Tech Notes 6038, Wash DC

Subject Index